Mister Zoo

The Life and Legacy of Dr. Charles Schroeder

The World-Famous San Diego Zoo and Wild Animal Park's Legendary Director

Douglas G. Myers
with
Lynda Rutledge Stephenson

Published by The Zoological Society of San Diego

All photographs are from the collections of the Zoological Society of San Diego, the Schroeder family, and Fotografia Praha. Photo credits: Ron Gordon Garrison, F.D. Schmidt, and Charlie Schroeder.

End page drawing by Charles Faust

Book and cover design by Arturo Lima

Printed in Korea

Sands Print Group, Ltd.

ISBN 0-911461-15-9

Library of Congress Catalog Card Number: 99-61695

Name *ZOONOOZ* Reg. U.S. Pat. Office

"He was Mister Zoo to us in the zoo world."

—Palmer Krantz, Riverbanks Zoo

"You may be Mister Zoo in San Diego, but to me and to your numerous friends in the scientific world, you are an outstanding scientist in the field of comparative medicine."

—Ross Nigrelli, New York Zoological Society

"Dear Mister Zoo. . ."

—Dr. Duane Rumbaugh, Georgia State University

"If ever there were a towering giant in the zoo world, it was Charlie Schroeder."
— Warren Thomas, former director, Los Angeles Zoo

"Charlie was a man who made everything live, made everyone want to belong."
—Leonard Goss, former director, Cleveland Zoo

"Dr. Charlie Schroeder will be always considered as the model director of a zoological garden."
—Ermanno Bronzini, former director, Rome Zoo

"Charlie was a go-go-go-guy; the success of our zoo is due to him. He was Messianic on zoos."
—Howard Chernoff, former San Diego Zoological Board member

"Charlie was sort of a con man. He knew more ways of getting money out of people than you would believe."
—Herbert Kunzel, former president, Aardvarks, Zoo supporter group

"During the nineteen years Dr. Schroeder was director of the Zoo, it grew in size and stature to the largest wild animal collection in the world."
—Ronald Reagan, former California governor and United States president

"The San Diego Wild Animal Park is a good example of the zoo of the future."
—Henri Hediger, former director, Zurich Zoo

"Few San Diegans have contributed to our city and region as significantly as Charlie Schroeder. Wherever a San Diegan travels in the world today, he or she is likely to be known as 'the person from the city with the wonderful zoo.'"
—Clair W. Burgener, former member, U.S. Congress

"The concept of the modern zoo owes a great deal to Charlie Schroeder's initiative and resoluteness. The conservation movement took on new dimensions under his hand."
—Ernst Lang, former director, Basel Zoo, Switzerland

"Charlie's enormous talents, boundless energies and dynamic promotional talents set a mark in the zoo world that few will ever match.
—A. F. Oeming, Alberta Game Farm, Canada

"No person or place touched by Charlie's personality was ever the same again. Somehow some of the Schroeder charisma rubbed off."
—James A. Oliver, former director, New York Aquarium

"Dr. Schroeder was a modern Noah...."
—Jim and Rose Mary Nastasia, photographer and artist

"The zoo plays a major role in our lives, it really does. Kids, especially, should come and see who they share this earth with."
—Dr. Charles Schroeder, in an interview with the *Strait Times,* Singapore

*This book is dedicated to all the people, present and past,
who have given their lives, their talents, and their unbridled love
to the work of the World-Famous San Diego Zoo
and Wild Animal Park.*

Acknowledgements

No one can tell the story of a man's life without the help of the hundreds of people who knew him, many who are still alive and many who have already left us, as Charlie has.

Through the course of our research, we were able to gather stories from many of Charlie's former employees, old friends, and long-gone contemporaries, sometimes in happy, crowded luncheons, sometimes in personal one-on-one interviews, sometimes in long distance faxes or recorded phone conversations, and quite often in yellowed letters and dusty oral histories.

Nothing would please me more than to name each and every one. Yet so many people shared their memories about Dr. Schroeder, more often than not with a laugh or a tear, that it would be impossible to name everyone. Even though I cannot name all of you, I want to thank each and every one who participated. You have played a special and personal part in the creation of this historical project.

There are a few, though, who must be mentioned for their phenomenal help in recording Charlie Schroeder's revolutionary Zoo director era.

Thanks to the San Diego Zoological Society's Board of Trustees for making the important decision to tell this pivotal period of the San Diego Zoo's life before it was lost, and to Dr. Kurt Benirschke, president of the Board, for envisioning the book's need and championing its cause. Thanks to our editor, Mike Yorkey, for his fine hands-on contribution. Thanks to Neil Morgan who opened his *San Diego Evening Tribune* column archives to us. Thanks to Chuck Bieler and Jo Hammershoy, who were touchstones for the Zoo's history and the book's research. Thanks to Peggy Blessing for her formidable organizational skills, and to Amy Pat Rigney and the volunteers who transcribed hundreds of pages of interviews. A special personal thanks to John Wexo, Dawn Dawson Wexo, and Maxine Schroeder for their incalculably important assistance.

And lastly, thanks to the members of the Zoological Society for their continued love and devotion to the animals for whom Charlie gave his life's work and continuing legacy.

CONTENTS

Prologue

Sorry, Dr. Schroeder

On a cool day in March 1970, San Diego Zoo photographer Ron Gordon Garrison and assistant Fred Schmidt were snapping photos and taking movies of giraffes being moved from the Zoo to the new Wild Animal Park being built out of 1,800 acres of high-desert dirt. They were driving through a small valley that would one day be the South African exhibit when they looked up and saw a figure in plaid shirt and khakis, marching up the hill carrying a load of stakes and a hammer. The man was Dr. Charles Schroeder, the director of the San Diego Zoo.

Ron and his assistant rushed up the hill toward the director, watching as he stopped, frowned and pounded a stake into the ground, then strode away to stop, frown and pound again.

"Dr. Schroeder!" Ron called to him. "What are you doing?"

"Laying out the monorail track," he yelled to the two huffing and puffing to reach him. "Everybody's taking too long to get to it."

"Well, stop for a minute, and we'll take your picture."

Never breaking stride, already eyeing where the next stake would go, Dr. Schroeder called back: "Don't have time! Just keep up with me."

There was never a more perfect metaphor for Charles Robbins Schroeder throughout his remarkable career.

With his ear-to-ear infectious grin, his speech habitually peppered with the words "fantastic" and "it works!" he was always seeing things others didn't see, plotting how to make them happen, and urging everyone he knew to join in and keep up.

Scientist, veterinarian, visionary, leader, taskmaster, mentor, P.T. Barnum in a suit and tie, Charlie had a remarkable mix of talents that fit perfectly for the moment in time he found himself director of a zoo with potential for uniqueness. And if others didn't see where he wanted to go and what he saw over the hill, that was okay. He would show them—and keep on showing them until they did see it.

We in San Diego knew him as the man who took the Zoo that Dr. Harry Wegeforth created and transformed it into an open-air botanical garden of endangered and rare species, naming it for what it was—the World-Famous San Diego Zoo. A hometown reporter wrote on his retirement in 1972: "The Zoo that visitors see today, where fences have tumbled and eye-pleasing moated areas have replaced many wire cages, is really The Zoo That Schroeder Built."

His largest legacy, though, may be The Park That Schroeder Built.

The San Diego Wild Animal Park, where the borders between a zoo and the wild blur, is truly the zoo of the future, a place as close to the wild as modernly possible, a captive breeding ground for zoo populations, and an ark for endangered species.

Around the world, his colleagues knew him as a pioneer in zoo pathology, the dean of worldwide zoological circle, and the visionary who built the Wild Animal Park to emulate and envy. By the time he retired, he was not only Mister Zoo in San Diego, he was Mister Zoo internationally, so much so that he spent his "retirement" traveling the world dispensing his zoological wisdom.

Entertainment. Education. Research. Conservation. These are the modern zoo's guiding principles. Early on, he was able to see how it all could be juggled—and juggled as close to perfection as Charlie Schroeder could make it. And oh, how that man could juggle. And oh, how he expected, and usually received, the same excellence from anyone standing nearby.

He was one of those rare men who found himself at the special moment in time with the right dream and the right mix of abilities to make the dream, beyond all odds, become real. Today, decades after his retirement and years after his death in 1990, he seems to be a larger-than-life figure as we reap the benefits of his hard-headed, full-hearted fight for the future—our future, the future of wild animals, and in many ways, the future of the "zoo" in general as a viable institution in our modern, wildlife-threatened, species-dwindling world.

"Make no little plans," goes the famous quote by turn-of-the-century American architect Daniel H. Burnham. "They have no magic to stir men's blood."

That might as well have been Charles Schroeder's quote because we have all felt the magic of his big plans.

In his introduction for *It Began With a Roar,* the story of the early San Diego Zoo and its creator Harry Wegeforth, Dr. Schroeder wrote this about Dr. Harry: "The world has known no more stubborn or determined man. As with many other successful builders, the impossible was just a little harder and took a little longer. He could enlist the help of a camel." Charlie might as well have been writing about himself. The parallels between the two men and the bulldog charm that made their dreams come true are unmistakable, as was their impact on the San Diego Zoo and the city they loved.

"There were some shortcomings and frank disasters, but all were overshadowed by his triumphs," Charlie wrote in that 1953 introduction. "It was a great day for San Diego when Doctor Harry chose our fair city for his home."

We could say the same for Doctor Charles, and we should.

As veteran *San Diego Union-Tribune* columnist and *It Began With a Roar* co-author Neil Morgan once wrote, Charlie Schroeder was one of the men who "built the San Diego we love."

Charlie would have chafed at having a book written about him. Friends, family members, longtime employees, and colleagues have all made that clear. He would have given a self-conscious shrug and said that most of the credit should go to others. He would have spoken about the team effort and how a good manager seeks invisibility. When a reporter tried later in life to get him to take credit for the Wild Animal Park, the old research scientist answered: "Fleming had an idea, but then he needed chemists, nutritionists, and so on to produce penicillin." He was right. But then again, to use his allusion, without Sir Alexander Fleming, we might not have had penicillin, at least not as quickly nor in the same way. Besides, there was no way that Charlie Schroeder with his 1,000-watt smile, arm-twisting gift for righteous causes, and memo-writing attention to details could be invisible for long.

But he loved a good story. All who knew him could find themselves suddenly on a walk through the Zoo or the Park being regaled by an interesting zoo tale or tidbit of information he'd just read. He'd get so animated that he'd have to stop and face his listener when telling the good parts. You never rode in the back seat when Dr. Schroeder was driving because he'd be looking back in full conversational mode instead of looking at the road.

The story of how Charlie became Dr. Schroeder and how Dr. Schroeder, veterinarian, found himself carrying an animal surgeon's dream into the future is certainly a good one. But this isn't just a biography of a man, even though such a biography would be a worthy read; few of us know anyone who's been knighted by the Czech government or bought endangered animals on just a handshake.

This is also the story of the modern San Diego Zoo and the visionary creation of the San Diego Wild Animal Park. Charlie Schroeder's story is a tale worth telling, one Charlie himself has told if all his walks and all his talks were strung together. He lived, breathed, dreamed, and plotted for the Zoo and for what it could mean not just for San Diego but for the world, wild and tame.

So his story is our story.

Marlin Perkins, another famous zoo director, once said that beyond all his accomplishments, Charlie was a jolly fellow to be around. Charlie must have loved that, especially at the end of his life. "Call me Charlie," he kept telling many of us around the Zoo following his retirement. But very few who worked for him over the years could address him as anything but Dr. Schroeder, not out of a sense of distance but out of a sense of respect for the history he created and the future he pioneered—for the story we are living today.

So, sorry, Dr. Schroeder.

It's time to tell the story you gave us to tell.

Section 1

CHARLES ROBBINS SCHROEDER, D.V.M.

Pioneer Zoo Vet

"What makes a zoo veterinarian? A general practitioner who has a keen interest in and is devoted to wildlife and who is willing to be a pioneer. He cannot hope to master the subject in his lifetime, but he can have a fabulous time trying."

—Charles Schroeder, D.V.M.

CHAPTER 1

Charlie's Beginnings

There's a famous story told about the young zoo veterinarian Charles Schroeder. On Christmas Eve in 1938, Charlie was working late at the Bronx Zoo's animal hospital when a member of the zoo staff passed by. Noticing the lights on at the rear of the building, he decided to wish Charlie a Merry Christmas. Walking through the building and out the back door, he found Charlie in his long, white lab coat, a white apparition against a blanket of snow that had been falling all day, performing a necropsy on a camel. He was softly singing, "Silent Night, Holy Night."

Someone once said that Charles Schroeder was a vanishing species—an eternal optimist. But Charlie began life in 1901, a time when optimism was America's theme. Any little boy, even a poor child from the tough side of the city, could surely grow up to make a difference, be it in politics or medicine or the zoological world. All he had to do was try hard enough. Charlie once explained the moment in time from his point of view: "Consider. We were just seeing the first automobile, the first electric light, the first telephone. And flying was unthinkable. The Bronx Zoo was just two years old. The Philadelphia Zoo, the next largest, was only twenty-five years old. The London Zoological Society was formed fifty years earlier. The San Diego Zoo didn't really open until 1922. Interest in the diseases of wild animals hadn't come into its own at all. The field of zoo medicine was a whole new discipline."

What forces made the turn-of-the-century boy born as Charles Robbins Schroeder into the man who would revolutionize the zoo world? The answer is a fascinating look at early influences that are as unusual as Charlie himself. Roller-skating through the Central Park Menagerie, swimming in the polluted East River, exposure to unorthodox schooling and social reform, a penchant for brawny adventure, the convergence of world politics, and the medical miracle of antibiotics, as well as a few wonderful coincidences, all played their parts.

Growing Up Tough

Charlie was born in Brooklyn, New York, in a neighborhood called Germantown, where wildlife could be found in the street outside your door. He would laugh about the fact that, at birth, he weighed an off-the-chart twelve pounds, six ounces. "I grew wider but not much longer in the intervening eighty years," he'd say, poking fun at his five-foot, six-inch height and stocky frame.

When he was four years old, his father was killed in a trolley accident and his mother, a violinist, had to turn to managing apartment buildings for a living. He grew up tough and tenacious in New York's Bedford-Stuyvesant area, where, as Charlie once said, a kid had to belong to a gang to survive. It was a place where street wars were common, with stones and bottles the ammunition and ash can covers the only defense, and where the election-night fires on street corners often ended in shootings. It was also a place where the neighborhood butcher gave him free chunks of bologna and all the liver he wanted for his cats, where an old German always gave him a banana, where hitching rides on the back of horse-drawn brewery trucks was entertainment, where "the English sparrows got their vitamin B complex from what the horses left behind," said Charlie, "and where the land was so flat that everybody had a stoop on the front of their buildings connected by hooks, and after every rain it would flood the streets so bad that everyone had to go running around the neighborhood to collect their stoops."

He learned to swim in New York's polluted East River, dodging garbage and raw sewage. That sounds unthinkable today, but as Charlie put it, "If you don't know about bacteria, you can't be afraid of it." But he was obviously a testimony to the maxim that what doesn't kill you makes you stronger because by the time he began working on a polio vaccine at Lederle Laboratories, he found out he was absolutely immune to the scourge of his times. "I was valuable," he loved to say. He always believed his good health, energy, and stamina came from being exposed and "vaccinated" for nearly every disease known to humanity the hard way—by taking in typhoid, polio, and other such deadly germs with each childhood swimming stroke.

As for schooling, his mother made an early choice for him that made an incredible difference in his life. She enrolled him in a new concept school called an "ethical culture school," which was controversial and unorthodox for its time. Felix Adler, its German creator, was a revolutionary educator. Not only did the school's teaching methods sound like those of today's progressive schools—studying the American Indian by making Native American food; learning about Eskimos by making candles out of blubber and even tasting the blubber as Eskimos might; bringing animals to the classroom to teach our relationship with the world's other creatures—but all the methods were based around the core idea of teaching ethics to schoolchildren, an idea that seems very contemporary, too. The school taught an early awareness of responsibility to humans and animals and the world. Charlie must have listened well because throughout his life he seemed to hold himself to some invisible standard of moral behavior and clear ideals that he never questioned—and that he expected from others.

Roller-Skating Through the Menagerie

As pivotal as that unusual schooling must have been, getting there may have been just as influential. To reach school he had to roller-skate from their apartment and travel across Central Park, and he liked taking a shortcut through the Central Park Menagerie, as the Park's zoo was called. One can only imagine what went through young Charlie's mind as he studied Native American and Eskimo life in school while rolling past captive animals that were from some of the same wild places. He once told a reporter that he and the animals became friends, and even then he hated the bars and wire confines holding them in.

Something, though, made a city boy interested in agriculture—maybe the same desire for freedom from the confines of the city he had felt for his "friends." After finishing high school in New York City during World War I, he earned a horticulture degree from the Institute of Agriculture at Farmingdale in Long Island, just about the same time as an animal-loving surgeon named Harry Wegeforth heard a lonely lion roar in an abandoned exposition of animals and decided San Diego should have a zoo.

But young Charles still hadn't made up his mind about a career. Finishing his degree studies, he worked with a group called the College Settlement, a group of women from Vassar and Smith colleges doing social welfare work in New York, running a summer camp and producing milk for the campers. After that experience, he thought he might like to try something away from the city, like dairying or farming. He didn't even think of veterinary medicine until he began studying agriculture, but then, instead of an interest in domestic animals, he realized his fascination was with wild animals. In old-fashioned terms recorded for an 1983 oral history project with San Diego State University, he explained that interest by saying, "Having studied agriculture breeds and breeding, and feeds and feeding by way of the Morris textbook reading, I developed a sincere interest in pursuing the study of the maintenance and the exhibition of wild animals."

To do that, he needed to become an animal doctor, so Charlie "pursued" the field of veterinary medicine by first studying the curriculum available at Cornell University, New York State College of Veterinary Medicine, then Colorado College at Fort Collins, followed by the College of Veterinary Medicine at Pullman, Washington.

This "education trail" was another way of saying he drove across the entire country to check out each of these different veterinary colleges personally. Getting there, obviously, was more than half the fun and most of the adventure.

Into the Twenties With a Roar

Just as the Roaring Twenties were beginning, Charlie saw America the old-fashioned way—by working his way across the vast continent with a series of odd jobs that spoke as much of the exciting period he lived as it did of Charlie's life. He and a couple of pals put a station wagon body on a 1917 Ford and followed the harvest as they headed west. In Kansas, they harvested wheat. Charlie's job was to straddle the back of a mechanical combine that was pulled by eight to ten horses and sew up the bags of grain as they were filled. In Denver, he worked on trolley cars, building snow sweepers while he investigated the veterinary college at Fort Collins.

Then Charlie heard that the vet school in Pullman, Washington, was offering such glowing enticements for students that he decided to head there as fast as the jobs would let him. After the wheat cropping and trolley work, he and his pals continued toward California and then up through Oregon, pitching hay, picking fruit, and performing other hired-hand jobs.

Finally in Pullman, he liked what he saw. He enrolled in the vet school, found a room at a "frat house"-style home packed with Masons, and stayed. While Charlie was twenty-four, the average age of the men in the house was thirty-four, many being World War I vets.

Dr. Myron Thom, one of his "frat" brothers, tells about a good-natured yet ox-strong Charlie Schroeder, vet student: "He was known for being an outstanding tumbler and gymnast who, despite mild manners and soft voice, could be a formidable opponent. His quiet demeanor usually made him the butt of the inevitable bully. This he would ignore as long as possible, then he would appear wordlessly stripped for action—thus activating second thoughts on the part of the offender."

Young Charlie, mannerly or not, must have loved those stripped-for-action moments. He seemed bent on finding the most challenging, burly, best-paying jobs because the one he found to pay for his veterinary education was a job he loved dearly and talked about the rest of his life. He worked several summers as an oiler for Looking Back Lines, a freighter ship company that would take him to the other side of the country again, but this time by way of water. In a Washington harbor that first summer, he had seen one specific boat that hauled lumber to the East Coast. It was a freighter that went through the Panama Canal and then up to New York City, Boston, and back to New York City before reversing course and returning through the Canal by summer's end. He boarded the boat while it was still loading lumber and asked for a job.

The engineer looked him up and down and said, "Meet me in San Francisco, and I'll take you on."

San Francisco? He had no money, but he had to beat the freighter there. Someone told him about a "coast-wise boat," a fast passenger ship that sailed from Seattle to San Francisco—and about a time-honored young man's way of paying for the trip that suited his stocky, strong-backed, adventure-loving sensibility just fine. In Seattle, he and a friend walked through the crew gangplank without anybody questioning them. The rest Charlie always explained with a twinkle in his eye:

"Then you crawled into the 'fireman's focsum,' the crawl space under the bunks, and stayed there until the whistle blew and the boat was out. Then you reported to the first engineer who was accustomed to this, who sends you in to get a bite to eat, then puts you to work cleaning the deck plates until you get to San Francisco. That's what we did. Then we made a beeline to the boat lines and got our job as oilers."

It was a job he loved each summer while he was a veterinary student. "I remember my last trip," he'd tell you, setting you up for a big exciting sea story. Then he'd say: "I had to throw all my neckties overboard because they got filled with bedbugs, and the food was pretty bad. But I liked the crew. We really had a ball."

California Dreaming

By 1929, he was ready to graduate, but to graduate in the same year as the country was plunging into the Great Depression was a perilous proposition. It surely must have given even a natural optimist like Charlie reason to pause, especially since he wanted to use his training in unusual ways.

And as for the normal path a veterinarian graduate followed, that world was about the care and nutrition of swine, cattle, goats, horses, chickens, turkeys—domestic animals not wild animals.

"The field of zoo medicine was a whole new discipline," Charlie explained, "and new disciplines always had a hard time gaining respect quickly. Nothing was known about the parasites, bacterial infections or the role of tuberculosis in zoos. Knowledge of the maintenance, nutrition, parasites and infectious diseases of captive wild animals was absolutely unknown."

Using one's education to pioneer the medical study of wild animals during a time when educated men were selling apples on street corners was not an easy dream. Few zoos around the world even employed full-time veterinarians. The New York Zoological Society, better known as the Bronx Zoo, was the first in the country to hire a zoo veterinarian. In San Diego, Dr. Harry Wegeforth took note and began working toward finagling a salary for a full-time vet. He accomplished it by getting San Diego County to pay the new veterinarian's salary, whoever he was to be, by offering the vet's services as the county's poultry pathologist.

But that lucky man wasn't Charlie Schroeder—not yet.

Dr. Wegeforth was great friends with Ellen Browning Scripps of the Scripps newspaper fortune. He wanted a hospital for the Zoo because he knew, as Charlie did, that there was much to learn about animals. Mrs. Scripps was willing to give $50,000 for such a hospital, but when Charlie finally arrived as the Zoo's second veterinarian several years after it was built, he found they had been forced to cut corners.

"The hospital had a couple of floor furnaces, sinks with no hot water, eighteen-foot ceilings, and no holding facilities for animals," Charlie once explained. "It was very pretty but cheaply constructed. They first hired a man from Ohio, a Dr. Whiting, to be 'research director.' But Dr. Harry failed to tell him that with this fancy directorship he didn't have any back-up, no lab help, no budget—and that the salary was paid by the county, so he would be doing lots of work with the poultry people bringing in their dead chickens to determine the cause of death. He had it after a bit and left."

Meanwhile, back in Charlie Schroeder's world, the new veterinarian was offered a job that must have seemed like a career for a lifetime.

The day of modern medicine was dawning, and suddenly veterinarians were needed in new medical research to study the nutrition of mice, rats, guinea pigs, rabbits, dogs, cats, and especially monkeys for the explosion of vaccines and wonder drugs created for humans and animals by new "pharmaceutical" companies.

Charles Schroeder, brand-new D.V.M., was offered such a job at the newly-established Lederle Laboratories in Pearl River, New York, later called American Cynamid Company. When he arrived, 250 employees worked there, and within only a handful of years, almost 30,000 workers were part of the production of penicillin and antibiotics such as aureomycin, and part of the research into poliomyelitis, tuberculosis, and rabies. Over the next forty years, he would work twice for Lederle Labs, once for the Bronx Zoo, and three times for the San Diego Zoo. Ultimately he found his calling and passion in the zoological world, but he often talked about his years at Lederle—three years in 1929-32 and thirteen years during and after World War II—and the thrill of being a scientist during the development of modern miracle drugs that we take for granted today.

Young Dr. Schroeder also met a woman named Margaret Wolfe at Lederle Laboratories during those early years. They fell in love, married, and later became parents of two children, Charles Randolph and Mary Ann.

Charlie might have spent his whole career in New York, except that one of his vet school professors visited the new San Diego Zoo.

CHAPTER 2

"I Got Your Boy"

Dr. John Howarth, who taught Charlie bacteriology, pathology and poultry diagnosis, met Dr. Harry Wegeforth and the Zoo's executive secretary Belle Benchley on a trip to San Diego. He saw first hand what was happening at the little upstart California zoo with the fancy new hospital—which needed a full-time vet capable of inspecting more than dead chickens.

Dr. Howarth told them, "I got your boy."

That "boy" was, of course, Dr. Charles Schroeder. Dr. Wegeforth and Belle Benchley offered him the job, and later Charlie would say of those early years, "I had a ball."

In a letter accepting the new position, he wrote as only a medical man would: "I have spent some time in the West and fear the good old California germ is still with me. In fact, I am a carrier and have transmitted the infection to several people, including Mrs. Schroeder. Being beneficial, it develops remarkable but pleasant symptoms."

He drove once again cross-country, but this time in a 1930 Model A with his young bride, Margaret. After miles and miles of sunny California expectations, they arrived in a rainstorm. To make things worse, the Zoo's dirt roads washed out with every rainfall. That soggy view was their first impression of a place that would capture his imagination and his passion for the next sixty years.

Zoo Vet and Everything Else

During his first stay with the Zoo from 1932-37, Charlie used to joke how his job covered everything. He'd say he was the veterinarian, pathologist, bacteriologist, virologist, serologist, grounds manager, clinician, research director, forage buyer, and poultry pathologist for the county. Oh, and he was also the self-appointed zoo photographer.

Belle Benchley, the executive secretary of the Zoo, fought the battle of the budget without end. Feeding the animals was even a struggle. When Charlie saw that there was money to be made in picture postcards, he roamed the grounds in the late afternoon with an old Graphlex that produced postcard-size photos. On weekends, he'd turn part of the research hospital into a black-and-white darkroom, sticking the wet photos on the building's metal incinerator where carcasses of dead animals were consumed. At night the pictures would dry, plop off, and fall on the floor. The next morning, volunteers would collect them, run them down to the front gate by the time the Zoo opened, and sell these original creations for much-needed pennies and nickels.

Conditions in those early years at the Zoo were, in Charlie's words, "quite primitive." You wouldn't want to ask Charlie how the Zoo fed their wild animals in the Depression years because he'd tell you: "I'd go to Mexico, buy old horses from Mexican skinners, dip and spray them, and bring them across the border. Then we used to stand the horse up, shoot him in the head, skin him, split the carcass, then the keepers would come in with their own knives and cut meat off the carcass for their animals."

Charlie knew what he was getting into. "Back then, being research director really meant you just learned as you went along," he'd say, with a big, knowing grin. "Dr. Harry, Belle Benchley, and

I would get together, discuss things, and do what we had to do to keep things alive. We didn't know how to prepare food, and we didn't know how bad some of the exhibits were." Dr. Harry had, very early, attempted to pioneer the concept of the moated enclosure. What he accomplished was amazing for his time, but there was never enough money. There were always cages worse for wear and animals who deserved better.

Of course, it was a different world back then. This was in a time when you found your animals in the wild, sometimes traveling yourself, and you thought nothing of it. Dr. Harry was famous for his long-distance travels in search of animals for his new zoo.

In his capacity as zoo veterinarian, Charlie would go on trips off the California coast with the ship *Velero* and Captain Allan Hancock to catch sea lions and elephant seals. Their numbers seemed endless, and these sea creatures could be traded with other zoos for animals they didn't have. Yet nobody would know what to feed the sea lions on their return or how to save them when they became sick. That fell to Dr. Schroeder, who would begin his first attempt at real zoo medicine research with these sea animals.

No "Tinker's Dam" Circus

Thankfully, research was as important to Dr. Harry as it was to Charlie. Wegeforth once wrote him: "My whole ambition was to have the Zoo revolve around the research. I do not a care a 'tinker's dam' to merely run a circus."

Yet both Dr. Schroeder and Dr. Wegeforth knew that since zoo medicine was in its infancy, far too little was known about exotic animals and far too many veterinarians knew less than they did. For dogs and cats, the research seemed to be expanding daily, but that was because there was money and interest for domestic pets. Everybody had a dog or a cat.

Dogs and cats, though, were only two species. "Think of all the other species in the world," Dr. Schroeder said whenever possible to veterinary colleagues. He believed Dr. Harry was right—that what they discovered about wild animals could also be good for humanity as well.

Since they both had their doubts about using veterinarians for the exotic problems of exotic animals, they instead called medical doctors in town. Did a monkey have an eye problem? Dr. Schroeder called an eye doctor. Dr. Harry talked his fellow physicians into coming to the Zoo to do studies, and payment was their names on a plaque somewhere on the grounds. Dr. Harry, Belle, and Charlie even started a research council, which was designed not only to help with research but, as Charlie admitted, "to get people sympathetic enough to give us things, such as a Mr. Burlingame, who owned a surgical supply company. He would give us a lot of cheap instruments and secondhand supplies."

Everything was new and anything went, if it helped the cause for the animals. By the time Charlie became Zoo director twenty-five years later, the San Diego Zoo had the first comparative virus laboratory in the world for domestic and wild animals—the Biological Research Institute.

But in 1932, he was on his own, sailing in uncharted territory. By all accounts, he loved the ride. His days were full of unusual tasks such as treating a bear with cataracts, force-feeding elephant seals, removing a broken bottle from an elephant's foot, treating the tail of a giant ant-eater, collecting saliva from wolves, and studying ulcers in sea lions. Charlie began writing and publishing papers on his discoveries, helping to foster the new discipline of zoo medicine. Ultimately, his San Diego work during the thirties, especially in animal pathology and emphasis on hygiene, cut the Zoo's death rate nearly in half. The zoological world noticed.

Marie, the Sniffling Walrus

One of Charlie's first projects and the subject of his first mark on the zoo medicine world was Marie, a little walrus that a philanthropist had brought to the Zoo. Back then, people gave the Zoo animals. It happened all the time. A sea captain had found the sick walrus, toted her home on his eight-knot freighter, and brought her to the Zoo to find out why she had the sniffles and other health problems. Marie was suffering from walrus skin lesions, was having difficulty breathing, and was being fed regular cow's milk by a bottle.

The more she was fed milk, though, the more her eyes would puff up, and the more her breathing would be labored. Of course there was nothing written on walruses' health at the time to help him, so Charlie called a medical doctor at the Scripps Clinic, explained little Marie's problems, and the doctor told him it sounded like an allergy. Charlie knew truth when he heard it, so he mixed clams and cream and fed that to her in the bottle.

Later, common walrus knowledge would be that walrus milk is rich in butterfat and that walruses are lactose-intolerant. That's why Charlie's mixture worked nicely, the cream being rich in butterfat, just like walrus milk. The lesions disappeared, as did the sniffles and the rest of the symptoms. Dr. Schroeder wrote up the case, and the article became the first time an allergy in an animal other than a human was ever described in a medical journal. Soon he had worked out a special formula and Marie, as well as the other lucky seals and walruses at the San Diego Zoo, thrived.

You Stupid, Stupid Man

But with Marie, he also learned a lesson in fund-raising from the "master" himself that lasted a zoo-filled lifetime.

As Charlie loved to explain it: "I wanted to find some money to build an exhibit for Marie with a little coolant for the water. So I went to La Jolla to see a very rich man, and he gave me $1,000. I was delighted. I found Wegeforth and said, 'Dr. Harry, I just got Mr. So-and-So to give us $1,000!' Know what Dr. Harry said to me? He said, 'You stupid, stupid man! That guy was worth $100,000 to us and you cut us off at $1,000!' "

Charlie learned a highly useful lesson about running a zoo that day, and it would not be the last one learned working for Dr. Harry Wegeforth.

CHAPTER 3

Lessons from Dr. Harry

Charlie continued to be taught by a true "master" about what could be done with innate talents and resourcefulness toward a good cause. Herbert Kunzel, former president of a Zoo supporter group called Aardvarks, once said in Charlie's presence that Charles R. Schroeder was "sort of a con man. He knows more ways of getting money out of people than you would believe." And no doubt, Dr. Schroeder howled at that because the same thing was eternally said about Harry Wegeforth.

"Watch out for this Wegeforth," John D. Spreckels, one of San Diego founding fathers, once said. "If you're a patient, you get your tonsils or your appendix out. But if you're working on the Zoo, you get cut off at the pockets."

Dr. Harry was legendary for some of the creative ways he'd get his Zoo what it needed. Like most people who knew Dr. Wegeforth, Charlie relished telling how Dr. Harry could talk money out of anyone for his animals.

Dr. Harry, he'd say, "borrowed" wood off railcars marked "WPA" for Work Projects Administration and told anyone who asked that the initials stood for White Pine Grade A.

Dr. Harry talked a bank into giving him enough money to put down a concrete foundation for a new enclosure without any collateral beyond his charm. He then raised the money to finish it during strategic strolls past the unfinished mess with his monied friends because it was much easier to sell a dream you could see.

Or Dr. Harry would mention, while walking with a visitor, that it was the Zoo's anniversary. He walked away with a check every time.

"The Zoo sure has had a lot of anniversaries this year," someone pointed out to Dr. Harry. His response? With a big laugh, he suggested the only way to handle that possibility was for donors to write checks for the real anniversary.

In *It Began With a Roar,* Neil Morgan writes of the day when a New York Zoo official was visiting Dr. Harry at the San Diego Zoo. At the right moment, Harry put on his best woe-is-me look and began moaning about the sad fact that no one ever gave him any turtles, which, by the way, were his favorite. "Could you send me some turtles?" he asked the important visitor from the big-shot zoo. The New York Zoo man agreed, walking to the Zoo's turtle collection to see how he might add to it for his friend Dr. Harry. Within minutes he had boomeranged back to Dr. Harry, half-shocked and half-amused, saying, "I have just visited your turtle basins, all forty-eight of them, you damn *rascal!* That's the biggest collection of turtles in the world!"

Obviously, Dr. Schroeder learned well at Dr. Harry's feet because the same "title" was bestowed on him half a century later. At Charlie's retirement banquet attended by more than 700 of his friends, former Cleveland Zoo Director Leonard Goss, who succeeded Charlie at the Bronx Zoo, stood up and said of the man of the hour: "We have heard a lot of accolades here tonight, but at heart he is a rascal."

Some of his old school chums could testify to his early rascal potential. Fellow student Myron Thom remembers the day before commencement when their dean noticed that someone had burned "Class of 1929" into one of the wood doors of the clinic. The dean called the whole vet class together and asked who had done the deed. No one stepped forward, so the dean announced: "Either it's removed by tomorrow or no one graduates." Thom remembered seeing Charlie Schroeder at 8 p.m. that same night, busy with sandpaper and paintbrush repairing the door. "To this day, I do not know if Charles was removing his handiwork or making sure he would graduate," Thom admitted.

That was one of Charlie's working virtues. Although you never questioned the man's integrity or pureness of motive, you also never quite knew what he was up to and what he might do. Even that famous smile always seemed to be saying he knew something the rest of us didn't, be it some hidden knowledge or the secret to squeezing the absolute most from life. It was probably why he could get away with as much as he did for the causes he championed throughout his life, and why so many people have rascal stories to tell about him.

Myron Thom's wife had a close encounter with Charlie Schroeder, Rascal Veterinarian, that may have helped her husband decide about that vet school door. For years, she had an uncontrollable fear of snakes that Charlie soon discovered. One day while they were visiting the Zoo, Charlie invited her to step into his office "to meet his new watchdog." To her shock and terror, she found an eight-foot boa loose on his desk.

"Pet him," Charlie ordered.

Frozen with horror yet strangely mesmerized, she obeyed. Amazingly, she found the sensation "not unpleasant" and never had an aversion to reptiles from that day on.

Then there are several accounts of Charlie's culinary way with a whole other species during the Depression years when a good steak dinner was a rare thing. Years after Charlie's send-off party, a colleague wrote him: "Who can forget the party when Lederle called you away? Everyone donated fifty cents for a suitable present and steak for the dinner. The present chosen took all the money, so beef couldn't be purchased, but you saw to it that we got beautiful tenderloins, sweet and tender! Surreptitiously and salubriously satisfied, Jack humorously emoted the nutritiousness of 'steaks equine versus steaks bovine' and all too suddenly, Ida was made aware that the currently digesting steak in her stomach was horse meat! She cried out, 'I'll never forgive you for this; I want my fifty cents back!'"

Either this was an ongoing joke of the man who just a few years earlier had helped butcher Mexican horses to feed the animals, or it truly was horse meat served up by the ever-pragmatic doctor of veterinary science. Either way, another friend tells the story about a 1940 picnic at which they devoured the most delicious steaks. When she thanked Charlie for the wonderful meal, Charlie's grinning response was: "It was horse meat."

Sheep Explosion

Some tales Charlie told a new generation of Zoo employees years later sounded like nothing more than old war stories, but they would have the strange habit of later coming back to life. Center for Reproduction of Endangered Species (CRES) deputy director of research Andy Phillips heard many of those stories. Charlie once told him about a strange situation that happened during his early days as the lone Zoo veterinarian. He had performed a necropsy on a Karakul sheep and had tossed the carcass in the incinerator to be burned. Then he started the incinerator and went upstairs to produce some postcards. Suddenly, he heard an explosion, then he smelled smoke. He rushed out of the darkroom to discover that the roof of the building was on fire and a huge flame was erupting from the incinerator's chimney. Apparently, the lanolin content of the sheep's wool was very high, and once the fat vaporized it exploded. The result was fire and flame blasting clear through the roof of the building.

Charlie said he and his crew put out the fire, but he didn't tell city officials for fear they would shut down the hospital. His solution was to quickly and quietly perform minor repairs to the roof.

"I just thought it was another story, like so many he told me," said Andy. "I never imagined it was true—until 1994." That year, the Zoo received a federal grant to refurbish the research building. One of the matching portions of the Society's contribution was to resurface the roof. One day the foreman came to Andy and told him that the roofing crew had found a major problem. In about 100 square feet around what used to be the incinerator chimney, all the huge support beams had been severely burned. "I couldn't believe my eyes," Andy said. "That 'sheep damage' had me begging fifty years later for another $20,000 to finally replace the beams!"

Look at Those Forearms!

Obviously Charlie Schroeder, D.V.M. was keeping his vet school, stripped-for-action physique. As a teenager, San Diego native Norman Roberts was always taking snakes to Si Perkins, the Zoo's reptile expert. Decades before Roberts grew up to chair a Zoo exploratory committee for a proposed Wild Animal Park, he was befriended by the new Zoo veterinarian Dr. Charles Schroeder. One day when Roberts had brought his latest reptile prize to the Zoo, he overheard Si and Charlie arguing in front of the reptile house. Then he heard Si yell, "I'm not arguing with you, Charlie. Look at those damn forearms!"

Charlie keeping himself in shape was a good idea considering some of the things that could happen to the only vet in a new zoo.

One day, he went to the airport to meet an incoming cassowary, a large flightless bird that has a three-inch nail on its middle toe. He wanted to give the bird some water, so he opened the crate to slide in a water dish. The big bird saw its chance to escape, crashing out of the crate and bowling over Dr. Schroeder—almost. Charlie wasn't going to lose the bird, so he managed to hook one of those forearms around the bird's neck as it passed. The cassowary lashed out with that three-inch toenail, slashing Charlie's pants from waist to cuff, then took off down the runway with Charlie running behind, clutching the bird with one hand and his trousers with the other.

Tiger Escape

As comical as such stories might be, some escaped animals are definitely life-threatening and a zoo's worst nightmare. In his book *Lifeboats to Ararat,* Sheldon Campbell, who grew up bringing reptiles to the Zoo and ultimately served as president of the Zoo's Board of Trustees, recounts a time when Charles Schroeder was at his most dramatically cool, zoo-vet brave.

One morning, minutes after the Zoo opened, a Bengal tiger—a cat capable of killing a person with a swipe of its paw—walked through a cage door left open by a careless keeper. He ambled up one of the Zoo paths, more on a stroll than an attempted escape. A stunned keeper saw the tiger and frantically called Charlie at the Zoo hospital. Automatically, Charlie ordered the Zoo closed and all visitors quickly and quietly evacuated. Then he called for all the available keepers to meet him near where the tiger was last seen. "No one had a rifle, so any effort to contain the tiger would have to be made without the comfort of a back-up bullet," Campbell wrote. "The stark truth was that the tiger had to be captured—had to be. Ticklish was not a strong enough word to describe the dangerous situation." Only nine keepers could be found to help.

The only safe way out of the situation was to drive the tiger back to its cage, Charlie told the men. He sent one keeper to flank the tiger, high on the hill above it, and to report any changes in the tiger's body language. No one was to corner the animal because the big cat might panic and charge. Charlie commanded everyone to form a line about ten yards apart, and then with any object they could, make as much noise as possible as they slowly advanced toward the tiger.

The tiger, who was still strolling along looking at the other animals, had gotten almost halfway to the six-foot barbed wire fence—a height any grown tiger could easily clear. The Zoo's perimeter fence was the only barrier between the tiger and the city.

"What happened next could have come right out of an old movie about a tiger hunt by the Maharajah and his guests in the days of British India," wrote Campbell. "The line of keepers, with Dr. Schroeder in the middle, moved slowly forward. Some beating sticks together, one had found two short lengths of pipe, another used a tweeter that he habitually carried to summon emergency help; others simply clapped their hands and shouted or whistled." The ruse worked. The cat froze. Campbell described the moment: "In the game of chicken that was being played, Dr. Schroeder, the keepers, and the tiger had one thing in common. Nobody wanted an eyeball-to-eyeball confrontation."As the sounds came closer to the tiger, the keeper-spotter reported that the tiger had turned and loped back up the road and into its cage, falling immediately into a striped heap, "as if it had just escaped the Maharajah and his hunters."

Legendary Energy

Mere mortals slow down as years pass and forearms lose muscle mass. Someone, though, didn't tell Charlie Schroeder.

Former director Chuck Bieler, who worked for Charlie and then succeeded him as director, always believed that Charlie "knew the secret to immortality." Maybe Charlie believed it, too, and perhaps that's the secret. Even at age seventy, Charlie could still get into his young, stripped-for-action attitude.

On tour of Africa as a retirement gift in 1972, he and his second wife, Maxine, were walking down a Nairobi street when a man ran toward them, grabbed his wife's purse, and kept on running. Instead of calling for help—the response of most sane tourists of any age or shape—Dr. Schroeder ran after the thief and tackled him on a rough cobblestone street. The *very* surprised purse snatcher, though, was able to wrestle free and run from this crazy, strong, and fearless "old" man.

And yet, that's not the last story about the "old" man.

On my first day of work as manager of the Wild Animal Park in 1982, I had my first close encounter with the fabled Charlie Schroeder energy. I pulled into the Wild Animal Park parking lot on a cold, rainy day. There stood Dr. Schroeder, wearing a Pendleton plaid shirt, brown raincoat, and his incredible smile.

He stuck out his hand and introduced himself. "I'm Charlie Schroeder. So, you're going to be running this place?"

"Yes," I said, knowing full well who he was.

"Well, I sort of had something to do with this place," he went on, "and I'd like to show you around."

"I'd love a tour," I said.

No, no, he decided. Since it was my first day, how about tomorrow morning? "I want to show you the most important thing in the Park."

"Great," I agreed. "Would 8:30 work?"

He gave me a look that said, *Hey, that's the middle of the day.*

I tried again. "How about 7:30?"

He smiled; 7:30 it would be.

The next morning, making a point to be early, I arrived at 7:15. The weather was even worse—rainy and cold. In fact, the weather was so bad that as I drove in, I thought—with great respect but not much knowledge of Charlie Schroeder—that the "old guy" might not be there. But when I turned into the parking lot, there stood Dr. Schroeder in the same spot, wearing the same brown raincoat and the same big smile.

Within minutes, we were standing in the valley where the Park's huge, solitary water line connects, and within another minute, I was straining to keep up with Dr. Schroeder as we strode down a semi-asphalted road overgrown with scrub bushes.

"I'm going to show you the most important thing of the whole Park," he said: "The water." We stopped in front of a fenced gate that was at least eight feet tall. He checked his pockets then looked at me. "You got a key?"

"I don't even have a key to my office yet," I answered, "much less a place like this."

The next thing I knew, he had climbed that fence and was gazing down at me from the top.

"Can you make it?" he said with an even bigger grin, and he hopped down on the other side.

He was eighty-two years old.

Big Heart and Warm Trouser Leg

Others could tell the same kind of stories about Charlie's big vet heart. He would always shake his head at how people would be so crazy over their pets. "Why do people kiss dogs? They're animals!" he once asked his wife Maxine, who bred golden retrievers. Yet he seemed a pushover for his friends' feelings about their animals.

Norm Roberts remembers the time when, as a boy, his dog hurt its leg badly. He knew only one vet in town with an X-ray machine, and that was Dr. Schroeder at the Zoo, so Norm carried his dog all the way to Dr. Schroeder at the Zoo's animal hospital. Charlie X-rayed the leg and cared for the dog as if he were one of the Zoo's irreplaceable animals.

Years later, Chuck Bieler would tell a similar story. He and his wife were heartbroken over their old, sick golden retriever, a dog they had received from Charlie's wife, Maxine, as a puppy. The dog was having convulsions, obviously dying. They called Charlie. As Chuck tells it: "Charlie said, 'Well, you should probably put him down. I tell you what. I'll come down there and I'll take care of it for you.' He came to our backyard and, with Judy and me there, gave the dog a shot and put it down. Then he said, 'Well, let me just take care of this, too.' He put our dog in the car, carried it to the Zoo, did an autopsy, and just took care of us all."

Then there's the story his co-workers at the Zoo hospital used to tell about the arrival of little Noel, a baby orangutan. Shy and frightened when she arrived, she was clutching a security blanket. She had to stay at the hospital for a short time, and while Charlie was getting her settled in, she grabbed onto his leg with her other arm. He walked around doing his work with Noel firmly attached. Not until much later, when her security blanket was sneaked away from her and she was forced to make a choice between Dr. Schroeder's leg and the blanket, did she finally let go.

Who Says I Like Everybody?

Dr. Charles Schroeder loved efficiency, cleanliness, and wild animals, but Charlie also loved people. His love for people seemed to go two ways, though. Those who worked for him said that he was easily enamored by people and was often disappointed in them, while at the same time adding that he never suffered fools gladly.

A close friend asked him, later in life, "How can you like everybody?"

"Who says I like everybody?" Actually, his philosophy was to give everyone he met a *chance* to be liked. He started with an open-arm attitude and then let them prove themselves unworthy of his trust. If they were fools, knaves, idiots, or loafers, they would eliminate themselves. If they showed initiative and talent, though, Dr. Charles Schroeder would be their friend for life.

He began mentoring early in his career, perhaps because he had wonderful mentors himself, including, most profoundly, Dr. Harry Wegeforth.

"I never worry whether something will succeed, only whether or not a person will succeed," Dr. Schroeder once told Robert Feeney, professor at the University of California at Davis, from which the Zoo gained many research assistants. Feeney remembered that quote throughout his career.

Back in the Depression years, many of the laboratory positions Dr. Schroeder offered weren't paid, but that didn't matter to most of the young scientists who wanted experience. While he expected everyone, paid or not, to do his best, he also seemed to know when to come down like a load of bricks or when to let the hard lesson be enough comment on the matter.

John Connor, a young assistant during that time, can attest to that—on two occasions. Dr. Schroeder had given him his first laboratory job, washing glassware and helping around the Zoo hospital. One day in 1934, a circus came to San Diego. Charlie agreed to remove an infected mammary gland from one of the camels. As John described it in a letter to Charlie: "My job was to keep the poor camel chloroformed, and apparently I did an excellent job because after you had tied off the last suture and stood up to survey the marvelous feat of surgery, it was discovered that the camel had gone to a better world, freed by an excess of chloroform. I will never forget how disappointed you were, although you never did chastise me then or later."

Then on another day, one of the Zoo's boa constrictors was brought into the hospital with severe mouth rot. John was instructed to pass a long glass tube down its esophagus and feed it warm milk. "On the first attempt, I got about fifteen inches of the tube down, and the poor animal somehow twisted and snapped the tube in two pieces, leaving one part somewhere in the esophagus and other part in my trembling hands," Connor wrote. "I'll never forget running frantically around the hospital, trailing about eight feet of sick boa constrictor behind me, hollering for you, half hoping to find you and half hoping not to find you. How kind you were! You took both me and the snake upstairs to surgery, made a small incision in one of the ventralfolds and esophagus, removed the glass rod, put in two or three sutures to close the wound, grinned, and told me to forget it."

Sometimes his mentoring could even mean a trip to the other side of the world—whether you wanted to go or not. Charlie, who understood the value of horizon-broadening experiences, never missed a chance to encourage others to travel, as one of his young research assistants found out during his tenure as the Bronx Zoo's veterinarian.

A collection of animals was going to be imported to the Bronx Zoo from Africa, but Dr. Schroeder believed the collection needed a scientist to travel to Africa to return with the animals, studying and caring for them. First, he convinced the Bronx Zoo's director, then he called in a young research assistant, Carlton Herman, who was struggling to make ends meet, and convinced him of this wonderful opportunity even though it would not pay a cent.

Charlie's final pitch was that at the very least, Carlton would be able to say the rest of his life, "When I was in Africa. . . ."

Dr. Herman, who later became the founder of the Wildlife Disease Association—in large part from discoveries gleaned from that trip—later wrote Charlie to tell him that he was right. "You'd be amused at the number of times your prophecy was fulfilled, every time with an image of your grinning face," he said.

Death Rate in Half

During those first years at the Zoo, from 1932 to 1937, Charlie made a remarkable difference as the San Diego Zoo's veterinarian. This part of his story is rarely told because his earlier life was overshadowed by his accomplishments as Zoo director. But in a time when detailed autopsies, an emphasis on hygiene, and thorough studies of animal physiology were considered new ideas in the zoological world, Dr. Schroeder was one of the first to instigate all three. The effect was immediate. As mentioned earlier, his extensive research program into the causes of animal death cut the Zoo's death rate by a stunning 47 percent.

He would become a pioneer in animal pathology, even though, true to his self-deprecating way, he never considered himself an expert. The results of his home-grown research, though, were undeniable. The zoological world soon took notice of this young, bright, energetic zoo vet. It wasn't long before he was offered a job at the New York Zoological Society.

CHAPTER 4

The Bronx Zoo/ Lederle Years

In 1937, Charlie moved back to the East Coast to become the veterinarian for the New York Zoological Society, or as most people referred to it, the Bronx Zoo. The job offered more money and the possibility of working in a much larger zoo context. Besides, it was the "creamy" place to be, as he put it.

"I wanted more experience, and, boy, did I get it," he would say years later.

The decision greatly upset Dr. Wegeforth, however. Setting the stage, even then, for Charlie's return—which Dr. Harry alone seemed to know would happen—he wrote Charlie: "Put as much time in with the up-to-date institution, and when you do come back, bring a wealth of knowledge. Your friend, Dr. Harry."

The only thing Charlie may have loved better than a new challenge was talking about what he learned, and when Charlie Schroeder described the exciting new thing he was learning, he would always sprinkle in "local color." Everything was interesting to him, and nothing was forgotten. He could regale you with stories of tapeworms or polio or black holes or whatever had happened to catch his immense curiosity at that moment. One of his employees at Lederle once said that he seemed "to be able to build up a head of enthusiastic steam over matters that escape the average eye or ear; a reflection of a remarkable and almost uncanny alertness that few people possess."

You could get Charlie talking about seals, for instance, and he'd be off telling you what seals taught them about viruses. It was all fascinating to this brilliant mind, and he could make it fascinating for the listener, too. That fascination would be a lifelong trait of a man who never lost the wonder of the world he lived in.

I saw it myself. Dr. Schroeder and I were part of a field trip group visiting a glacier during an International Union of Directors of Zoological Gardens (IUDZG) Conference after Dr. Schroeder's retirement. As we shivered and listened to the guide describe the purity of glacial water, I looked around to see eighty-three-year-old Dr. Schroeder on his knees at the water's edge, tasting melted glacial water. "Dr. Schroeder," I said, "can I help you?"

He looked up and made a face. He was assessing the purity of the water himself. "Not bad. Have you ever tasted the water of the Dead Sea?" he said, pulling himself up. "Very, very salty."

Many people remember Charlie telling several anecdotes about his years at the Bronx Zoo that were revealingly "local color" for the times. For instance, he'd talk of the animal dealers still in New York City in the late thirties, especially in the Bowery section. One family named Ruhe from Hamburg, Germany, had been buying and selling for generations, and they specialized in exotic birds—hundreds of them, thousands, he'd say, often ones that the zoo professionals or the Bowery dealers had never seen before.

One of the curators at the Bronx Zoo would actually go down with his textbooks and identify them for the dealers. Such was the abundant and unknown state of wildlife at that time. Charlie would talk about such animal dealers in his usual fascination, but it was a fascination tinged with disapproval, due perhaps to Dr. Wegeforth's efforts to band together with other zoo directors and trade among themselves, thus eliminating price-gouging dealers.

Another favorite story was about a New York Zoological Society Board member during the twenties who rushed into the zoo from his tandem, a sulky drawn by two horses, and asked if he could have some help with a letter he needed typed. The reptile curator, the only one in the office at the time, told the man he had a young lady who would be pleased to take his letter.

"Is she gainfully employed?" asked the trustee.

"Of course," the curator answered.

"If she is gainfully employed then she is no lady," the man sniffed.

Decades later, Charlie would enjoy telling that story after watching the changes for women's equality in the workplace, especially as it affected his zoological world. After retirement, he'd regularly show up at the Wild Animal Park to have lunch with the employees, many of whom, by that time, were women keepers. He would come away deeply impressed with them all.

"In 1932, we wouldn't have had women keepers on a bet," he explained in delighted amazement about the Zoo's early days. "We'd have thought having male and female keepers in the same area would have made all sorts of 'moral' problems. But now, they take the whole thing for granted. I wouldn't have been able to imagine some of the young women we had lunch with today out there with a shovel and a pitchfork picking up manure and throwing it on a truck. They take it in stride and do very well." You can almost see him shake his head at this point. "And so educated!" he went on. "In the thirties, I remember at least two of the keepers signed their name with an 'X.' Today most of the keepers, especially the incoming ones, have Bachelor's degrees, and many of them Master's, and many have in sight a Ph.D. That was unheard of!"

Another Bronx anecdote he loved to tell on himself showed his still-intact, man-of-action bent, even when it offered the danger of fire-walking. One day, a rare Saiga antelope came to the hospital for removal of an abscess. The animal didn't survive the surgery. So, as he did with all animals that died, Dr. Schroeder promptly disposed of the carcass by putting it into a primitive two-story incinerator: the fire was on the first level under a grate, and the material to be burned was piled up on the second level. He started up the fire, then went to report the animal's death to Dr. Reed Blair, the Bronx Zoo director.

"What did you do with the carcass?" demanded the director.

When Charlie told him, Dr. Blair was livid. The antelope, on its death, had been promised to the American Museum of Natural History, where a study was being done on the Bering Sea bridge to prove or disprove the migration across the Strait. Dr. Blair had given his word, so Charlie ran back to the incinerator, opened up the second level, climbed into the hole, plowed through the gases and smoke, dragged out the semi-charred antelope carcass, and lived to tell about it again and again.

Then there was the time the director happened to drop into the basement of the hospital to find a Charles Schroeder personal research project, created on the principle that it is easier to ask forgiveness than permission—a principle Charlie held in high regard.

"What's this?" Dr. Blair asked.

"It's a steer," Charlie dutifully answered.

"What's it doing here?" the director asked.

"Well," Charlie began, "this is a cooperative venture with Mt. Sinai Hospital's cancer research unit, wherein we built a box stall here in the basement in order to study the. . . ."

"Get it out," ordered Dr. Blair.

What was it like to be at the Bronx Zoo for Charlie? Working in a place steeped in tradition was sometimes a problem for the budding visionary. He soon found out that he was ahead of his colleagues in many ways, even in this "creamy" place. He began to change things the moment he arrived.

William Bridges in his book *A Gathering of Animals: An Unconventional History of the New York Zoological Society,* mentioned the flurry of activity surrounding the arrival of a new veterinarian named Charles Schroeder:

> "Dynamic" is the term that must likewise be applied to a new member of the park's staff. Dr. Charles R. Schroeder, who succeeded Dr. Noback as veterinarian, came back to his native New York from the San Diego Zoo, where he had been veterinarian for the previous five years, and immediately started to modernize the Animal Hospital and reorganize its functions with energy and enthusiasm seldom seem around the Zoological Park since the earliest Hornaday years. He set up an elaborate record-keeping system, organized a group of distinguished

medical men to do research in cooperation with the Animal Hospital, charmed the executive committee into providing money for modern laboratory equipment, and proclaimed as his goal the practice of preventive veterinary medicine, with all that implied in the way of quarantine, close supervision of exhibition quarters, better diet, early recognition of disease, and so on.

Panda Tapeworm Immortality

One of the highlights of his New York years was the chance to work with the first giant panda ever to be seen in this country. Her name was Pandora, and she went on exhibit at the 1939 World's Fair in New York's Flushing Meadows. Charlie enjoyed talking about how playful the panda was and how the keeper and the panda would wrestle away the greater part of the day.

One of Charlie's duties was to figure out what to feed it. He discovered the panda did very well on sugarcane imported from Louisiana, along with bamboo from the New York Botanical Garden.

Working with Pandora also offered him his one claim to scientific immortality. A species of roundworm he discovered in the panda is named after him: *Ascaris schroederi* is its scientific title—a fact that always struck Charlie as humorous.

The lasting work he did there, though, was self-instigated veterinary detective work. In the thirties, great quantities of rhesus monkeys were being imported for polio research. One of the suppliers for the Zoo was also one of the principal suppliers for the research. During one casual conversation, he told Charlie he and most of the other suppliers were losing monkeys in droves from tuberculosis. This piqued Dr. Schroeder's curiosity and concern.

"I thought it would be interesting to learn the source of the TB," he said. Working with suppliers in India, he discovered the interesting fact that there was no TB in the field. In other words, the monkeys were catching tuberculosis after being captured. But how? He kept investigating and found out about a strange custom. Most Indians spit at such animals. That simple act, all by itself, would transmit human tuberculosis to the monkeys. When the monkeys were jammed into crates for the long boat trip to research hospitals, the monkeys would spread the TB among themselves. At that time, the current research test for tuberculosis was one hundred years old. Using his budding gift for bringing talented people together and inspiring them to a good cause, Charlie helped pioneer a TB screening test for the monkeys, then demanded improved shipping techniques with bigger crates, fewer animals, and better food.

Because of Charlie's curiosity and cleverness, thousands of monkeys were saved, which ultimately helped polio research. The new screening test for primates became standard usage for decades, and it also furthered the study of treatment of tuberculosis in humans.

Intellectual Ferment

For a man with tireless curiosity, the new ideas whirling through the intellectual climate in New York at that time may have made more of an impact on Charlie Schroeder's future worldview than anything else.

The New York Academy of Sciences was not far from where he worked, and he talked often of the lectures and the exciting scientific community he found there. War was brewing around the world, and New York was becoming a haven for many well-known scientists from Europe and Asia. It was a time of incredible intellectual ferment, and Charlie must have relished every second.

He spoke of having lunch with Sir Alexander Fleming, the discoverer of penicillin. One of the curators of the Bronx Zoo, William Beebe, could be called one of the first ecologists, and Charlie counted him as a prized colleague and teacher. Eunice Minor, the executive director of the New York Academy of Sciences, became a close friend. Soon Charlie was a member of the biology section of the Academy and would continue the connection during his thirteen years at nearby Lederle Laboratories.

Being exposed to so much stimulating intellectual scientific thought would affect his own thinking for the rest of his career. Marlin Perkins, the famed St. Louis Zoo director and TV's *Wild Kingdom* host, would say that Charlie had an encyclopedic knowledge of veterinary medicine. But the truth was he had an encylopedic knowledge of many subjects. Since Charlie had the knack of knowing a good idea when he heard it and the uncanny ability to file it away for future use, this was no doubt a heady time for him.

What Are You Still Doing Here?

In 1939, Dr. Harry Wegeforth, who had traveled to New York to pick up some turtles from the Bronx Zoo, strode into Charlie's office. "What are you still doing here?" he asked good-naturedly. "You got what you came for. Now come back." And then he repeated the same message to the Bronx Zoo's director that very afternoon: "I'm going to take Charlie away."

Charlie promptly returned to San Diego.

"I am glad to see you come back to a zoo that is young, for I feel you will be able to see it in full bloom before you pass out of the picture," Dr. Harry wrote to Charlie on his return.

Charlie thought he *would* stay until he passed out of the picture, but he would remain only two years in San Diego. Charlie would later say he was there long enough to help build several Zoo WPA (Work Projects Administration) projects—such as the warehouse with the Zoo's first refrigerator units and the rebuilding of several of Dr. Harry's first attempts at moated enclosures—before the problems of the world intervened. Unfortunately, the United States would plunge into a global war, and his talents would be needed for the war effort.

War Effort

In 1941, Lederle Labs, already preparing for the United States' impending entrance into the war, asked Dr. Schroeder to return to help in the creation and preparation of new medicines and vaccines. Charlie didn't want to leave San Diego, but he did. Charlie did not mention the coming war effort in his resignation letter, however. Interestingly, the young family man mentioned that his reason for leaving was "purely economic."

"I have been offered a position in a large corporation that gives some guarantee of progressive increase in income. I feel that I cannot legitimately ask for or expect an increase in salary here until the wage scale is increased for the other members of the staff, which apparently is not possible at this time." And then he ended with these words: "The thought of leaving San Diego and the Zoo is not pleasant."

Breakthrough Scientist

Charlie rejoined the Lederle staff as manager of their animal industry section, and later became director of Lederle's veterinary clinical research. With Stanford and John Hopkins University, Lederle produced a whole host of medicines used to treat war wound infections that could be used very quickly by both the United States Army and Navy.

That was just the beginning of his important contributions to modern medicine. For the next thirteen years, Dr. Schroeder met and worked with a group of scientists who were playing pioneering roles in a cure for lobar pneumonia and in the development of antibiotics such as aureomycin, modified virus vaccines for polio, rabies, distemper in dogs, equine encephalomyelitis, and a great array of vitamins. He played an integral part in developing a diphtheria antitoxin for horses, and his lab became the first in the county to provide antitoxins to protect horses from human diseases.

At Lederle, having learned the major effects of hygiene in his early San Diego Zoo vet years, he was quickly known as "a housekeeper." Charlie liked things neat and tidy, and what he liked became gospel in any lab under his eye. Years later, a Lederle associate would write that her memory of Charlie was of a young man "full of enthusiasm, drive, and with a passion for cleanliness in our environment."

He also would continue to push the study of mortality rates, urging his profession to chase the knowledge he found in those studies during his San Diego years. He must have done it in a persistent "bully pulpit" way as only Charlie Schroeder could because decades later in honor of Charlie's retirement, a one-time colleague would write this glowing tribute to Charlie's persistence: "Some day we will have a workable system of morbidity and mortality reporting in animal diseases. When that happens, I hope the record will show that Charles Schroeder was the fellow whose enthusiasm and confidence held his colleagues' feet to the fire until the job was done."

In many ways, the experiences he had during that period of life was a dress rehearsal for his coming zoological leadership years. He wrote more scientific papers than ever during that time. He became chairman of the biology section of the New York Academy of Sciences, which he had loved a great deal during his Bronx Zoo years. He used to talk about the tiny lunch room for the faculty and research people and the great discussions he had over meals, which made it "better than a faculty club."

Many new breakthroughs were happening in Lederle's labs. Every day, patent attorneys would walk into the labs and ask, "Schroeder, what've you got today?" Charlie would tell them what everyone was doing, and they'd decide whether the work needed a patent.

The skills he had been honing over the years, both in research and managing people, were put to the test. This was a time for excellence, a time for nothing but one's best effort for a greater cause. No one appreciated that more than Charlie.

When he was promoted to head of veterinary production at Lederle, a man who thought he was in line for the job made Charlie's life difficult. He was disruptive at every meeting and thuddingly slow and surly with his own work. This went on for several weeks until one day Charlie told him, "Okay, come on. We're going to accounting."

"Why?" the man asked.

"Because we're going to write your last check."

From that time on, he had no problems with anyone. He expected the best from the best, and that included himself. He expected you to show your talent, work up to your potential, and give it your all. If you did that, then Charlie Schroeder, Boss, would give you room to shine and freedom to fail. If you gave less than your best, then it would be time to say goodbye. Quickly.

It would be a style of management he would bring back to the San Diego Zoo to cheers and jeers, and not a small amount of fears.

CHAPTER 5

Dr. Harry and Dr. Charlie

Harry Wegeforth died in 1941, not long after Charlie returned to Lederle. Once Charlie told a close friend, John Wexo, that he consciously modeled himself after Dr. Harry. "He admired the hell out of him," said John. "Harry Wegeforth is who Charlie wanted to become." If that was his goal, then Charlie Schroeder succeeded. The parallels between the two men are uncanny and fascinating to explore.

Read these excerpts from the introduction that Charlie wrote for *It Began With a Roar:*

> I have never met a man with Doctor Harry's animal-like persistence, or his tenacity of purpose to do the job and carry it to completion, come hell or high water.
>
> —
>
> If it is selfish to make personal sacrifices in fortune and energy for the sole reward of seeing a dream come to fruition, then Doctor Harry was a very selfish man.

He knew how to pull the most out of everyone who would help, and you felt good about it. He gave you the authority to act in your position, as well as his support, and always his personal gratitude.

—

The world has known no more stubborn or determined man. Like so many other great public builders, his critics are endless. If you knew him and worked with him, you couldn't think of lying down on the job. His enthusiasm was catching. You knew that when you were on Doctor Harry's side you were on the winning side, and felt secure.

—

As many other successful builders, the impossible was just a little harder and took a little longer.

—

In a modern industrial sense, he was the perfect supervisor. He knew how to handle problems. He always let you know how you were getting along.

—

He gave credit where it was due. He would let you in on the plans and especially those that would affect you. He knew how to make the best of each person's ability.

—

Research in nutritional, parasitic and infectious disease of wild animals held a high place on his agenda.

—

Countless young men and women were inspired to become scientists under Doctor Harry's behind the scenes guidance: physicians, veterinarians, zoologists and researchers in biological fields.

—

There were some shortcomings and frank disasters, but all were overshadowed by his triumphs.

For those of us who knew Charles Schroeder as a friend, colleague, mentor, boss, or emeritus inspiration, the same words could have been written to describe himself.

Dr. Harry saw himself in Charlie, no doubt. Belle Benchley must have seen the resemblance, too, because she would not let him forget for a moment that the San Diego Zoo was where she thought he should be.

He kept up a running dialogue with Mrs. Benchley or better put, Belle kept up a dialogue with him—perhaps because the war had depleted her pool of reliable help, but more importantly because she recognized his immense zoological talents and knew his deep affection for the Zoo. Charlie was coming back if she had anything to say about it.

Dr. Schroeder and Mrs. Benchley

Charlie would laugh years later that he always seemed to be on call for Belle Benchley. "Will you go meet somebody on an airplane for us?" Belle would write him. "Will you give your comments on this matter?" She wanted to keep him in the "family" so badly that she offered him a "honorary" position if he would continue to give opinions and act as a sort of "free agent" for the Zoo. The concept was a good one because Charlie went for it:

"Dear Mrs. Benchley," he wrote in 1942. "Answering your note and expressing an opinion is a tiny bit embarrassing in my present capacity. I welcome some sort of honorary appointment and would work hard and long at this end for the Zoo, and in such a capacity I would feel privileged to make rather frank statements when I was called upon to do so."

From there, he gave his frank opinions and denouncements — quoting both an Eli Lilly circular and Isaac Walton's *The Complete Angler*—on a veterinarian's proposed practice of restraining confined and sick wild animals that Belle obviously found repugnant. This type of postal communication became common between San Diego and New York, at least for the next few years.

The war years were a difficult time for the Zoo, as it was for the country. Sometimes Belle would write to discuss day-to-day issues, such as employee and food shortages, worries she needed to share. "With poor and indifferent help in the warehouse and so many changes, it is not always possible to have food delivered just as it should be,"she'd write. While we have no absolutely definite food poisoning, there have been days when I supposed there might be."

Usually, the dialogue was a flurry of telegrams and air mail missives that demanded Charlie's attention:

Letter:
Mar. 4, 1944
Dear Dr. Schroeder:

I thought that if I sent you some air mail stamps you might find time to write more quickly about some of these questions. Thanks a lot for the help you always are to me. Mrs. Belle J. Benchley

Postcard:

Karl arrived Los Angeles Saturday. Off ship today. Elephants have edema at navel like MGM elephant did. Please wire me collect any information you have regarding the condition. All looks bad. Belle J. Benchley.

Telegram:

ANEMIA CAUSED BY HOOKWORMS RESPONSIBLE EDEMA GIVE IRON SALTS CONTACT HERMAN FOR MGM EXPERIENCE. Dr. Schroeder

Postcard:

Have wired Ruhe for price, sex on pygmy hippo, described in World Telegram. Please inspect, wire me via Western Union condition. Mrs. B. Benchley

Telegram:

MALE FIFTEEN POUNDS SMALL, APPARENTLY GOOD CONDITION, TAKING SOME SOLID FOOD IN HANDS. GAMBLE FOR YOU. IF INTERESTED AND PRICE RIGHT WAIT MONTH OR MORE UNTIL ON SOLID FOOD BEFORE SHIPPING. HEINZE IN PHILADELPHIA. TWO FINE RETICULATED GIRAFFE. Schroeder.

Telegram:

GET PRICE ON GIRAFFE TRY TO BUY PIGMY HIPPO NOT MORE THAN 1000 DOLLARS. Belle J. Benchley

As for Charlie's personal opinion of Belle Benchley, he loved and respected her immensely, but she drove him crazy feeding bon bons to her beloved Mbongo and Ngagi, the Zoo's famous gorillas. She loved to sit with them in their cage and feed them candy, and the gorillas became quite obese, much to Charlie's chagrin.

He admired her formidable talents, however, and her growing savvy reputation as a trader, He was also impressed by her own fascinating life story. In the twenties, she was a divorced woman in her early forties with a son to support. Taking a temporary bookkeeping job at the Zoo, the only kind of job open to women in that era, she turned her gift for animals and numbers—and cussing like a man when the situation deemed the need—into Dr. Harry's answer for his managerial problems.

Dr. Schroeder was most impressed, though, with her uncanny ability to sense health problems by even a casual glance at one of the Zoo's residents. She would stop by to mention that she noticed, while driving through the grounds, that "the Malaysian porcupine is ill," and he would find out she was right. She could see, even from the window of her car, the tiny changes in a sick animal before even he could.

Belle's best-selling memoir, *Life in a Man-Made Jungle,* would also expose the world to the San Diego Zoo and jump-start its image to the moment years later when Zoo director Charles Schroeder would add the words "World-Famous" as part of the Zoo's full and permanent name.

Charlie always hated the fact that, seemingly due to the times more than other factors, Belle never held the title of "director" of the Zoo, remaining "executive secretary" throughout her long tenure. The moment she retired and Dr. Schroeder was hired, one of his first executive decisions was to give her the title of director emeritus.

First, though, Belle had to make sure Charlie came back where he belonged.

CHAPTER 6

Belle's Pick

During his last few years at Lederle, Charlie became, among his other duties, a spokesman for the products he was helping produce, even traveling as far as Spain to lecture on them.

One day in 1952, while in Los Angeles on a public relations tour for Lederle, he made a last-minute swing to San Diego that began his next—and last—cross-country move.

Howard Keddie, host of the *Home on the Ranch* radio show in San Diego, received a call from an ad agency in L.A. The woman on the line said she represented Lederle Labs, which had recently developed an antibiotic for livestock and were anxious to let farmers know about it. Oh, by the way, their rep happened to be in L.A. If Keddie was interested, they'd be glad to send him down to be on the radio program.

The rep's name was Charlie Schroeder. After the broadcast, they had lunch together. During their conversation, Charlie told

Howard that he had once been a vet at the Zoo, and he asked Keddie if he would mind detouring by the Zoo so he could say hello to his old boss Belle Benchley before he left town. Howard agreed.

Years later, after Charlie was Zoo director and their paths crossed again, Charlie told Howard that he had no intention of coming to San Diego on that trip, but it was that quick hello that started the negotiations for Charlie's coming back to the Zoo as director.

"He used to love kidding me by giving me all the credit for his being here," Howard said.

Yet the story was no doubt true because very shortly after that visit, Belle, who was already in her seventies, told the board that if they ever wanted to see her retire, they better offer the job to Charlie Schroeder.

In 1953, Charlie picked up the phone at Lederle and said hello to an old friend, Robert Sullivan, a member of the San Diego Zoo's Board of Trustees. "Charlie," he said, "will you come back and be the director of the Zoo?"

Lederle was becoming a large organization. Charlie admitted to himself that he wasn't the happiest guy in the world, even though he had risen to a directorship there. He tried not to show too much enthusiasm on the phone, but he was delighted at the possibility. Charlie and Margaret's two children had been born in San Diego, and he had to admit that Belle Benchley was right; part of him had never left the Zoo. When he went home that night, he asked Margaret what she thought. She said a definite yes—she would go at the drop of a hat.

That hat dropped rather slowly, though, and the reason why speaks volumes about Charlie's character and desire to honor his predecessor. In fact, it would be a full year before Charlie arrived in San Diego.

At that time, Dick Yale was the printer commissioned to create the first edition of *It Began With a Roar* with Neil Morgan. He and Neil had access to all the Zoo's files, and during their research, they found out the Board had made arrangements for Charlie to come months before. They naturally wondered why he hadn't arrived. Whatever the reason, they both thought asking the new Zoo director to write the introduction for the book would be an excellent idea, especially since he knew Dr. Wegeforth so well. Why not ask?

Neil called to see if Charlie would write the introduction, and while talking to him, casually brought up the mystery. "By the way," Neil said, "I understand all has been ready for you to come for months. Why aren't you already here?"

"I'm waiting for Mrs. Benchley to retire. As soon as she does, I'll come," Charlie answered.

But," Neil informed him, "Mrs. Benchley has said she's waiting for you to finish a project *there.*"

Charlie laughed loud and hardy, then said he'd been finished for a year. He was packed and waiting for her to give the word.

Neil mentioned the story in his column and within days, Belle finally announced her retirement.

That day, Belle sat down and penned a letter to Charlie. "To know I am leaving this in such capable hands, to one who has so complete a picture of what Dr. Harry has tried to do makes me deeply happy," she wrote.

Another letter from an old associate of Dr. Schroeder, Waldo Schmidtt of the Smithsonian Institute, written a lifetime later would capture the moment well: "It was Mrs. Benchley who was so devoted to the Zoo and its future, that she could not rest or retire until she returned you to her great love and yours, the San Diego Zoo," wrote Schmidtt. "And now when we look about, we have before us the most impressive evidence of how many of her and Dr. Harry's hopes for the Zoo were brought to fruition, and carried forward beyond all expectations by you, Dr. C. R. Schroeder."

Beyond all expectations?

Probably not Charlie's because San Diego Zoo's new director ended his 1953 *It Began With a Roar* preface with these prophetic words: "There will be many improvements in construction and services, new specimens and novel exhibits in the years to come. The Wegeforth ball has been rolling under the expert guidance of Belle Benchley, and who knows how far-reaching the San Diego influence will be!"

Section 2

DR. CHARLES SCHROEDER

Inspired Zoo Director

"A zoo should be more than just looking at elephants. It should give you a closer look at the fascinating things in nature. When you look at an animal, you should say, 'What a gorgeous creature. Why does it have horns? Why is it that color? Why does it have cloven hooves?' Each animal species has a story to tell."

—Charles Schroeder, San Diego Zoo Director

CHAPTER 7

Schroeder Snowflakes

Kakowet the pygmy chimp always gave him away. The chimp's hooting and howling, begun the moment he spied the Zoo director each day, turned quickly into an early warning system for everyone in earshot. It meant that Dr. Schroeder, armed with his little black book and a pair of eyes that missed nothing, was walking the Zoo grounds.

His nightly walks became legend because they struck terror in the hearts of every employee in Dr. Schroeder's path. Shirttails checked, smiles in place, trash disposed, enclosures tidy—every imperfection had to be caught. Nothing escaped Charlie's practiced eye as he strolled, taking notes on hedges needing clipping, signs needing updating, and cages needing attention. All week long, he'd compile his black book's list, then on Sundays, he'd fire up his Dictaphone and dictate memos to his staff in triplicate. They called them "Schroeder snowflakes"—those little white pieces of paper—and called receiving them the "Monday morning blizzard." Ignore the memo the first time and you'd get a reminder of it. Ignore it the second time, and you did so at your employment peril.

Ask any Zoo employee who worked during those years of the most lasting memory of their boss, and invariably the story most told is about Charlie Schroeder's little black memo book and these solitary walks. These "lone prowls," as one journalist and friend called them, soon became the linchpin of his legend.

Charlie knew it. His favorite quip about his daily memo habit was this: One day he saw very dirty glass inside an enclosure, so Charlie found the keeper and pointed to the offending glass.

"How long does it take you to put a little juice on your windshield and wipe it off?" Charlie asked the man, pointing to the smudged window. "For God's sake, Joe, can't you see that?"

"Doc," the keeper said, squinting at the window, "if I could see that, I'd be the director."

The San Diego Zoo had to be perfect every day for every visitor, and it took everyone every day to keep it Schroeder-perfect. So if the "Joe's" of the Zoo couldn't "see" it all, the director would by keeping his eye peeled, his little black memo book and his Dictaphone handy, and the Monday morning blizzards of Schroeder snowflakes flurrying.

The habit must have begun the day he arrived. According to Charlie, the first thing he noticed was that the Zoo wasn't clean enough, especially on the weekends when no cleaning crew worked. The first purchases he authorized were for a mechanical sweeper with a steam cleaner to clean the Zoo grounds and a Dictaphone to create those memos to keep the grounds clean.

Between those two purchases, Dr. Schroeder began transforming the appearance of the Zoo immediately.

Dream Coming True

As delighted as he was about becoming the director of the Zoo, he told Belle Benchley he never aspired to it, perhaps because he could never quite see anyone replacing Dr. Harry Wegeforth or Belle, Dr. Harry's hand-picked manager. From his accomplished position at Lederle Laboratories, accepting the position must have looked like a crazy idea. Ironically enough, for a man who was still going and blowing as Mister Zoo until he died weeks shy of his ninetieth birthday, one of the reservations voiced about his taking the job was his age. He was fifty-three at the time. His response to the Zoo Board's Robert Sullivan concerning the offered job was this: "How many years am I good for? Will you get an active normal cooperative person who will efficiently and effectively manage Zoo affairs? To the best of my knowledge and our Medical Department, I am in the best of health, including pumper and no ulcers." This from a man who would later jet to Singapore and Iran in his eighties to offer his zoo wisdom.

The Zoo's interest in him was obvious. As he wrote to Belle Benchley about the job, "I have many friends in San Diego, in other Zoological Parks, and in scientific circles. I know Washington and its bureaus. I have just become a fellow and life member of the New York Zoological Society and I am a fellow and life member of the New York Academy of Sciences. My limitations are known to you!"

Yet he had just been given new responsibilities at Lederle as director of Veterinary Clinical Research and manager of Animal Industry Production. To come to San Diego would mean a pay cut and a loss in his early retirement, only two years away. He was an influential man from a very large company doing worthwhile work. Why did he leave?

When Charlie had said yes to the Zoo offer, Belle Benchley's response was, "I couldn't be more happy if it were my own son." She had carried on Dr. Harry's dream long enough to pass it on to a man she no doubt saw as the next generation's Wegeforth. It could only have helped that Dr. Harry's son, Milton Wegeforth, was one of the Board members instrumental in Charlie's possible return. A newspaper story grasped the poetic symmetry this way: "But the new director does not have his eyes entirely on the future. Like Mrs. Benchley, he thinks of the Zoo in terms of the plans and spirit of Dr. Wegeforth. He recalled Dr. Harry was always ready to try something new."

Maybe that last recollection of Dr. Harry helped Charlie's decision—that he'd be coming to a place packed with potential that he could wrap his formidable energy around. Or better, as one reporter put it, "Dr. Charles Schroeder, new director of San Diego Zoo, gives the impression of being a man for whom dreams are coming true." Those early days with Wegeforth truly were still very much with him. As he said back then, it was not as if he didn't know what he was getting into.

"In our business," he wrote Mrs. Benchley, "we feel that a supervisor's success is dependent 90 percent on his ability to get along with people and 10 percent on his knowledge of the job. I do know the Zoo and its many problems." And no one doubted his way with people.

He couldn't resist . . . again.

Moving Stone Tubs

What kind of director was the city getting? A story from his Bronx Zoo days was quoted as the example of what San Diego could expect of their new Zoo director. Lee S. Crandall, general curator at the Bronx Zoo, was asked by an enterprising 1953 *San Diego Evening Tribune* reporter to comment on Dr. Schroeder's Zoo director "potential." He answered with this memory:

"In the days when Charlie Schroeder was veterinarian at the New York Zoological Park, I went to visit him at his home. Getting no answer to rings and banging, I went prowling and finally peered down a cellar areaway. A muffled voice bade me come down. I came upon a curious sight," explained Crandall. "Charlie, in his shirt sleeves, was engaged single-handed in moving a slate tub weighing several hundred pounds. His children needed a sand box in the yard and Charlie meant that they should have it. The fact that moving the tub up the stairway was a Herculean task made no difference. With an auto jack and wooden blocks, he already had inched it halfway up. Horrified by Charlie's suggestions that I might like to help, I postponed my urgent business until the next day when Charlie reported his children as happily playing in their new sandbox.

"That's the sort of man Charles Schroeder is," said Crandall. "He copes mentally with problems of veterinary medicine or business administration with the same energy that he devotes to physical struggles with stone tubs."

Electric Eel or Pay

Charlie set out, just as he did with the sweeper and the Dictaphone, to make changes that needed to be made even if they seemed immovable as a stone tub. The Zoo that Dr. Schroeder inherited already had in place some remarkable people: Chuck Shaw with reptiles, K. C. Lint with birds, and George Pournelle with mammals. Belle Benchley had guided the Zoo into a world collection.

But its main battle was money, as it had always been. Charlie knew firsthand what running a zoo on a shoe-string, hand-out, donation-basis was like. He remembered watching Dr. Harry stop cars and ask for money to keep the Zoo operating, and he'd only have to tell you once about skinning those Mexican horses on the Zoo grounds to feed the animals for you to remember it, too.

A newspaper article that summer of 1953 showed the problem had not changed and was about to escalate. "Get Electric Eel or Pay, Zoo Told" was the headline. A city councilman discovered that the Zoo had been getting a free ride from the city for electric service. No doubt it had been that way ever since Dr. Harry's persuasive days. Belle Benchley's response to paying the yearly bill was that the $6,000 annual tab might mean the Zoo wouldn't be able to buy their next rhinoceros.

The witty city councilman's response was: "Perhaps we should advise the Zoo to acquire an electric eel."

This was the constant state of the Zoo's finances throughout its existence. Things had to change, and after years of acquiring business experience to match his scientific knowledge, Charlie must have felt he was prepared to help the Zoo make the change.

The only problem was Belle Benchley's last piece of advice. "Never treat the Zoo like a business," she told him.

"But this is a business," he later explained in his series of oral history interviews. Charlie saw that much of the financial problems that had plagued the Zoo's history could be fixed by simply running it like a business. "Belle loved her animals. No question about it, and I loved them, too," Charlie said. "But it was a new era." So yes, the Zoo would become a business, a very unusual business, but a business that could do amazing things if it had the money. "I instituted all the things you would have in a business right away," he said. "It's a continuous thing: money, money, money. You have to be very cold, very practical." If there were anything that Charles Schroeder had learned at Lederle, it was how to run a business working with animals for a good cause—and making money at the same time.

A profile in *California Veterinarian* magazine put it this way: "He is first an academic professional, and a medical man with a special affinity for research medicine. However, in his present position as director of a multi-million dollar institution, he also needs the administrative genius of a corporation president, the hard fiscal eye of a Secretary of the Treasury, and the public relations expertise of a Presidential press secretary."

Most men might be good at one or two of these traits; Charlie Schroeder amazingly had the gifts to do them all—and to love every minute of it. This was his chance to flex every muscle God gave him. In perfect Charlie Schroeder style, he hit the ground running.

The Long Way

Dr. and Mrs. Schroeder packed their belongings and took the long way to San Diego. They visited twenty-eight zoos in their trip across America, studying their operations. He filled in a battered notebook with pages divided into subject headings that covered everything from elephant buying to floor scrubbing.

There was no "rule book" for operating a zoo, he decided, so he would write his own. And as soon as he hit town, he was making news and talking about his notebook of ideas. A local reporter began her story saying it was "impossible to escape his enthusiasm, his drive and direction."

What was his stated direction in 1953?

Nothing much. Dr. Schroeder just wanted to make the San Diego Zoo the greatest in the world.

Plans? Oh, he had a few. Quite a few. And he gave every reporter who asked a taste in his vivid Charles R. Schroeder, over-the-top style. Why, the possibilities are endless, he must have told each interviewer, adding new ideas with every new story. Listen to a few excerpts:

> Our vision is to more effectively show the animals. First, we want to get a collection of rarities. . . .
>
> —
>
> We want to develop the medical department for the express purpose of properly treating and caring for the animals, and using preventive procedures.

> We want to build enclosures without wire. You can show the animals very effectively with nothing between you and the animal. That's the way it should be. Wire and bars will be discarded for open grottos whenever practical.
>
> —
>
> Ape mesas surrounded by moats might be practical, but might not do with gibbons, because they can jump too far. But certainly if the moat were designed right, we could do it for the gorillas. . . .
>
> —
>
> How about using windows in the big bird cages so spectators would not have to look between meshes of a wire fence? Maybe we could put the windows in a darkened room on the outside of the cage.
>
> —
>
> Would San Diego people like tractor trains better than the present zoo buses? They could be roofless.
>
> —
>
> There's a new-type rubber floor for cages that provides a softer surface, similar to the jungle floor, easier on their feet I'd like to see installed. . . .
>
> —
>
> There should be better methods of labeling exhibits. Also, it would be good to get more San Diego scientists actively interested in Zoo research.

From the very first newspaper accounts, he started shaking things up.

A Room With a View

Belle Benchley left on her trip around the world, a present from the Zoo's Board for her long and popular service to the Zoo, but she didn't clean out her desk. No matter. Dr. Schroeder set up his desk on a tabletop in the old Board room, and went to work.

When he finally did get an office, he found out rather quickly that the only restroom in the administration building opened into his office. Of course, in typical Charlie Schroeder style, he put a positive spin on his "room with a view."

"I got to meet the staff very quickly," he said. "It created a friendly atmosphere, as every-one nodded and smiled as they filed in and out of my office during the day."

Out of Short Pants

Charlie quickly noted that the Zoo had no organizational chart, no job descriptions, and no salary administration, so he gave titles to the appropriate men and upgraded salaries. Then, as had always been his habit, he drew on the resources of the community. With help of professors and graduate students at San Diego State College, he created the Zoo's first organizational chart.

"A good many years ago, the Zoo grew out of short pants and became a business large enough to require special consideration for equal pay for equal work in this metropolitan area for all Zoo employees," he wrote in the in-house zoo newsletter, *The Zoo Bell,* in announcing the changes, beginning with the change everyone would certainly like most.

The Zoo would now be conducted like a business, one that served animals as well as people, but as a business. The change was so dramatic that employees began calling the Zoo the "Cyanamid Subsidiary" after Lederle's parent company, American Cyanamid.

The annual operation budget was $500,000 when he arrived in 1953; when he left in 1972, it was more than $8 million.

But that's just talking money, and money was not the point. The point was what the money could do for the Zoo. Before he retired, the Zoo would be virtually rebuilt out of its own earned income, and as the income grew, so did the Zoo's sights for the future.

"The zoo world, just like academia and some other aspects of society—can have its ivory-tower types," stated a national magazine profile of Dr. Schroeder. "He was always very cognizant of the fact the Zoo's pioneering efforts depended upon the financial support the average visitor provided—that he had to appeal to a lot of people in order to be a financial success and therefore carry out the less visible programs that were necessary. He did a good job of balancing those diverse requirements," said the article.

That was a good assessment of the man. Charlie liked the whole idea of "pioneering"—a trait that would have people calling him a visionary by the time the Wild Animal Park became a reality. His first pioneering experience, in the twenties, had been influencing a new veterinary discipline for exotic animals that focused on properly treating and caring for animals using preventive procedures, and it was still a goal of his in the fifties.

Charlie sensed the role of zoos was about to change in a big way. In the fifties, zoo directors were more aware of the budding problem of endangered species than academic scientists. They were seeing it firsthand due to high prices for some species, and other species were unavailable at any price. From a pragmatic zoo perspective, animals die and collections must be replenished.

How to do that? That was the question coming faster than anyone could imagine.

And that would spawn another even more dramatic question:

To what end?

What role should zoos play in the world of limited resources that was coming faster than anyone could have predicted?

Charlie had to make the San Diego Zoo the place to come to again and again and again in order to have the funds to pioneer. Charlie very quickly would be pegged in print as an "accomplished vet regarded as a consummate business manager with P.T. Barnum-like flair of showmanship" because he was about to start juggling those skills, all at the same time to make it happen.

Instinctively he knew at least one thing he could do that people wanted, something he'd already begun.

The man known as a housekeeper at Lederle ("I liked things neat and tidy") was going to make the Zoo Schroeder-clean.

Neat and Tidy

"There is no reason for having a dirty zoo or an odorous zoo," Dr. Schroeder would often say. And that went double for flies.

He had been to zoos all over the world, from Russia to Rome, and they were always untidy. "They didn't see it," Charlie'd say in wonder. One year, he was invited to speak to Italian zoo keepers a few days before an international zoo meeting at the Rome Zoo. "They admitted to me that because of our international group's upcoming meeting, the zoo's director was running around with his people trying to do a last-minute scrub-up job because it was so bad," he said. "By the time of the meeting, overall it looked pretty good, even though behind the scenes you could see where they swept some of the stuff." When he happened to return to the Rome Zoo a week later, it was right back where it was before.

That was the "nature of the beast," wasn't it? Just the way things were at a zoo, any zoo, in the fifties.

Not to Charles Schroeder.

His first purchases as Zoo director, after all, had been the mechanical sweeper and Dictaphone for his inspection memos. The Zoo, he reminded employees, should never look like a trash-strewn carnival. He had the cleaning crew working full steam ahead—every day.

"I remember the first time I ever walked around the Zoo with him," said Bill Seaton, the Zoo's publicity manager during much of Charlie's era. "I was talking to him, and he was listening. Then he saw a broken branch thirty feet above us, scribbled 'fix this branch' in his little book, while never missing a beat in the conversation. He didn't stop talking a few steps later when he picked up a popcorn box and threw it in a trash can."

Charlie's view was this: "The people who come at 3:30 p.m. are entitled to the same clean restrooms as those who come in at 9 a.m."

He even hired a "sanitarian," a retired Navy man, to keep the Zoo restrooms clean and presentable. Like most topics he felt passionate about, he could—and often did—talk in great detail about the Zoo's restrooms.

"We redesigned every restroom," he said. "Wall-hung fixtures, a trough the length of the building under the fixtures, double-end entries so there can be no blind spots for anyone to hide in the back. Epoxy-covered walls so lipstick and other stuff could be wiped right off. Terrazzo or epoxy-covered floors that you could rinse off. . . ."

During those early years, he'd take employees on field trips to a new amusement park up the road called Disneyland, which kept its grounds spotless. Trash never seemed to stay on the ground more than ten seconds. Cleanliness may be next to godliness, as the old saying goes, but to Charlie Schroeder, they seemed to run neck and neck.

He had a gift for bringing people together that was so remarkable it would become perhaps his most famous talent. "He was the kind of man who made you want to belong," one of his colleagues would say about him, and this was his very first chance to do so, with a very basic, simple idea—a clean Zoo everyone could be proud of.

He wrote a small article for the Zoo's in-house newsletter called, "Who Has the Most Important Job at the Zoo?" In it, he made a case that the groundskeeper was just as important as the director, for without him the place would fall apart. Then Charlie came full circle with the thought so that by the article's end, everyone's job was the most important. "I wanted to remind everybody, that by George, my job's important, too," he said. "You couldn't operate without the keeper, the assistant keeper, the supervisor, the curator, the people who keep the records. Without everyone, where would we be?"

Shirttails and Dump Trucks

At age sixteen, Andy Phillips began working at the Zoo in the sanitation department, assigned to the trash detail. At that time the Zoo had its own trash truck, a Garwood Load Packer with a Chevy 409 dual-quad engine—the kind of engine the Beach Boys wrote a song about and something a teenager like Andy would consider a waste inside a trash truck.

"Charlie had suggested his director of personnel hire several retired Navy personnel, probably chief petty officers, whose job it would be to keep high school kids like me under control," Andy remembered. "Every day, Charlie would walk around the Zoo grounds and jot down any problems in his little black book. We all knew it. Charlie noticed everything, and no day went by when something wasn't correctable. Whatever problems he saw were passed on to the 'chiefs,' who, in turn, told us to correct them.

"One morning my boss came up to me, and I figured that I had done something wrong," Andy went on. "But instead, he announced the Load Packer was broken down. The plan was to have me stand in the bed of the Zoo's big dump truck while others threw the trash cans up to me. I would then empty the cans in the bed of the truck, smash the trash with my feet, and pass the empty cans back to those on the ground."

That day, he remembered, was particularly hot. "The Zoo was going to be packed with people, which meant packed with more trash. By mid-afternoon, after spending several hours dumping cans of half-eaten hamburgers and fries, as well as everything else, the smell of the dump truck was almost overwhelming. I had untucked my shirt in an attempt to keep a bit cooler," Andy said. "We had just stopped in front of the Safari Kitchen for another load of trash. About halfway through the passing back and forth of twenty trash cans, I noticed Charlie Schroeder's bald head staring up at me. Here I was covered with ketchup, mustard, and grease, was surrounded by flies, and what does he say?

"'Phillips, tuck your shirt in,' he said. 'We don't want the public to think we run a sloppy operation.'

"He then wrote something in his black book and walked on. I hadn't realized he even knew my name."

How Do You Know If a Zoo's Good?

Cleanliness became so much part of the San Diego legacy that when he retired, he was asked to write a memo on how to keep it Schroeder-perfect.

> INTEROFFICE MEMO—CLEANLINESS
> To: Charles Bieler
> From: Dr. Schroeder
> July 26, 1973
>
> It was my habit to attempt to get in to the Zoo at 4:00 or 4:30 each evening and with notebook in hand record such small items as soft drink cans in bushes off the beaten trail, holes or cracks in pavement, cracked concrete, rusted expanded metal on rails, fresh carvings on trees, battered, rusty, dirty automotive equipment, dirty windshields, plantings in need of pruning, water or removal.
>
> One has to have a trained eye to see evidence of bad housekeeping, and everyone is not equipped to recognize untidiness.
>
> Attention must be paid to the physical equipment in restrooms, paint, broken door latches, broken or nonfunctioning light fixtures, damaged plumbing.
>
> Restrooms should always be immaculate.
>
> Exhibit signs must be kept clean and must be updated identifying the animal in the enclosures.

All exhibit areas should be continuously occupied—never an empty cage.

Births and special exhibits should be announced by special signing.

Avoid temporary pencil and crayon signs.

Attention should be paid to food waste—or inadequate spoiled food—too few stations—unclean water and containers.

Personal tidiness should be observed, too. All should abide by the grooming standards set by the personnel department—hair, beards, mustaches, shoes and working attire are all important.

Policing the ground by any and all employees to prevent, control or report abuses to the animals, to our plantings, to labeling is essential.

Grounds tidiness unfortunately requires hard labor.

The list is endless and maintenance crews will never quite catch up.

This is not all inclusive but indicates the typical problems of daily concerns.

CRS

"How do you know if a zoo is a good zoo?" Ed Green, a fellow Rotarian, once asked his friend Charlie Schroeder long after he had retired.

His answer?

"The first thing you do is go to the men's bathroom," Charlie said. "If it's clean, that's how they treat their animals. If it's dirty, that's how they treat their animals."

Today, few of us can walk the Zoo grounds without picking up trash and scanning for untidiness in the time-honored Schroeder way. It's an attitude, Charlie seemed to be saying, an attitude about what we were all trying to achieve here in this special place in Balboa Park.

CHAPTER 8

The First Big Idea

With the Zoo on its way to being the cleanest zoo in the world within weeks of his arrival, Charlie's launched into his first big idea—a Children's Zoo. He'd seen children's areas at the Bronx Zoo and the Catskill Game Farm. Under zoo supervision, kids fed and played with small animals. Families loved the special, set-aside areas, and Charlie knew why. The concept helped satisfy what he believed to be the public's desire to mingle with animals, especially familiar ones. While visitors wanted to see more exotic animals, they also loved areas stocked with lots of animals they could touch and feed and love up-close.

Charlie knew just how to get this dream to come true.

Once again, he learned well from Dr. Wegeforth. He knew of Dr. Harry's ingenious arrangement for the Zoo to be held in trust by the city "for the children of San Diego," an arrangement that saved the Zoo many times in its formative years from becoming a political Ping-Pong. Anyone eyeing Dr. Harry's Zoo for political gain or even, sometimes, for unpaid electric bills, found themselves in a fight with the children of San Diego.

A proposal to build a Children's Zoo section seemed more than appropriate for a Zoo that was by its charter in existence for San Diego's children. Plus, the strength of public opinion in San Diego, when it came to the Zoo, was a formidable thing, and Charlie knew it.

So in 1954, wasting no time, the new Zoo director finagled the question of a Children's Zoo onto a Chamber of Commerce poll. The result: 69 out of 100 San Diegans favored a Children's Zoo and said they would take the children there to pet the animals.

"SCHROEDER, SHREWD POLITICO" was the headline for the story the next day.

With poll ammunition in hand, he began to petition the Board of Trustees. And then, in the style he'd exercise the rest of his career, he didn't wait for anything so pedestrian as a vote to get the ball rolling. He went right ahead working out details, getting everything in place, creating plans and illustrations—all designed to let the Board see what a wonderful concept their vote could make real. The experience would be a trial run for the "run" of his career, the Wild Animal Park.

He had even found the perfect place for the extension. Balboa Park's two-acre Japanese Village had been razed during World War II, so Charlie, acting for the Zoo, acquired the land and targeted it for the Children's Zoo, certain he could persuade the Board to build. "The Bronx Zoo's was one-third of an acre, so ours should be about two acres," Charlie said later, no doubt with a sly grin on his face.

Then he gathered up several young architects, along with a man named Chuck Faust (who would become the Zoo's chief designer), and asked them to begin brainstorming every Wednesday night about a Children's Zoo.

As Charlie had done through the years, he would hire talented people, then he would give them the freedom to see what they could do, becoming, as always, just as much as cheerleader as a demanding boss. Chuck Faust, an ex-Air Force pilot and artist, may have been the epitome of that sort of mentoring.

What excited Charlie about this young designer beyond his ideas was his explicit drawings of new concepts. Charlie, who had seen "visual aids" work with Harry Wegeforth, saw firsthand the power of persuading people by showing them something to get excited about. Dr. Harry would show prospective donors a half-finished exhibit and the money would roll in; Charlie found that a drawing could work the same magic, if done right, and Chuck Faust could draw. That dynamic alone probably kept Faust's phone ringing with Charlie's latest idea to "make real" one of his percolating ideas.

San Diego's Children's Zoo had to be unique. He told Faust, let's do it special, do it different, and do it right.

"Everything that Dr. Schroeder ever had anything to do with, he put his whole heart into it," said Faust. Charlie expected nothing less from everyone else.

By 1956, even while his new Children's Zoo design "staff" continued to work, the Children's Zoo had still not been given the okay by the Board. The detractors were not silent. Mingle kids with baby animals? That could be a disaster! And where would the money come from?

The logic of these arguments fell on deaf Schroeder ears. He'd seen children's zoos work elsewhere; it'd work here.

He had firm allies on the Board, such as president Milton Wegeforth and Howard Chernoff, but he could not quite get the final majority's approval.

Finally, he decided to get some help from public opinion.

He talked to Neil Morgan, city columnist for the *San Diego Evening Tribune* and, not coincidentally, co-author of *It Began With a Roar.* In May 1956, the following memo found its way into Neil's column: "From C.R. Schroeder, DVM, managing director, San Diego Zoo: 'We know the children's zoo is going to materialize and I'm sure it will be a great success and a real feather in San Diego's hat, but your support will just bring it about that much faster.' "

Public support happened. Within weeks, the Board voted yes and the money materialized. The project cost $200,000 and the entire cost was covered by private donations. In the summer of 1957, the first child bought a ticket from a child-sized ticket booth, then whizzed down a slide to enter The Paddock. Debby the guanaco, Kim the llama, and Mercurochrome the alpaca were waiting to be petted, fed, and loved—in a safe environment.

Everyone held their breath.

Finally, this entry appeared in Neil Morgan's column: "Some skeptics viewed with alarm the plan for intermingling tots with animals and birds in the new Children's Zoo. But with almost a month of operation past, there hasn't been even a bandage-sized emergency. Worst casualty so far is that of a father who leaned over a grotto to pet a bear and lost his upper plate."

Soon the papers were pointing out that the grownups were outnumbering the tykes at the new Children's Zoo.

Do Something Different

Chuck Faust and his architects tried things with the Children's Zoo that had never been tried before. The ones that worked were copied by zoos around the world.

"I was surrounded by a neat bunch of people, like Dr. Pournelle and Chuck Shaw," remembered Faust. "They were all very anxious to do something different, so everybody was behind just about everything that we did."

Before they were finished, they would rebuild the Zoo from scratch, and many ideas began with the experimentation inside the Children's Zoo. The best example was a walk-in aviary, which became quite an innovation in the zoo world.

Faust one day suggested that kids should be allowed to walk through the birds' cage. "The old Zoo crowd said, 'Oh, that's never going to work. The kids are going to get in there and pull the wings off the birds,'" he remembered.

Keepers were also leery. Who could blame them for worrying? This was something no one had ever tried. Could children be trusted to walk through without harassing birds or letting them escape?

Faust left it to Charlie: *Can we do it?*

Yes, Dr. Schroeder said. *We'll try it.*

So Chuck and his group built two walk-in aviaries—a quail exhibit and a finch exhibit—both with two sets of double swinging doors and lots of free-flight room inside. Children walked in, oohed and ahhed, giggled, and screamed in delight as special birds fluttered and flitted around them. It all worked beautifully.

The two walk-in cages were such a great success that they quickly moved the idea to the regular Zoo. The large flight cage built in 1928 was revamped and became the first walk-in aviary in the world. The idea was this: Build different levels with waterfalls. Eliminate barriers. Add an occasional gentle rain.

The experience would be like being immersed in jungle instead of just walking around it. The concept worked beautifully and zoos around the world followed suit. We've continued to refine this wonderful experience today in both the Zoo and the Wild Animal Park.

Steaks and Logs

One day when they were building the Children's Zoo, Charlie decided they needed some children-sized benches. He asked Chuck Faust what to do about it. "Why don't we get some logs, shave off the top and use them for the benches?" answered Faust.

Dr. Schroeder found out about a sawing crew up on the back side of the Cuyamaca Mountains sixty miles away. He called Faust and said, "Meet me out in front of the Zoo at 5 a.m. in the morning. We're going to up to the mountains." Faust met him the next morning, and they climbed in the Zoo's pickup truck, which was twenty years old and "pretty shot," according to Faust. Charlie insisted on driving.

When they got to the mountains, Charlie turned onto a two-lane dirt road with a big hump in the middle, which they scraped most of the time even with the big, high truck. They traveled about twelve miles on that dusty road.

"Finally, right before we got to the saw camp," Chuck remembered, "he suddenly stopped the truck, turned the engine off, and said, 'Come on. Get out, get out.'"

Dr. Schroeder walked to the back of the truck and pulled out a box of food, kicked some of the dirt away, and to Chuck's pleasant surprise, started a fire and was soon cooking steaks on a little grill he'd brought.

"I couldn't believe my eyes. It really went over big with me, though, I can tell you that," Chuck added, chuckling. "We had a great meal, then we went up to the saw camp. They sliced off some logs for us, then loaded them onto the truck, and the truck's springs went *adios,* but we made it back."

Charlie returned Faust to the Zoo well-fed and with a taste of what the years ahead would be like working for Charlie Schroeder.

The adventure was just beginning for the Zoo as well.

Quality of Zoo Life

Throughout the Zoo, the sudden emphasis was on quality of life and quality of sight. Chuck Faust was what one reporter called "a man dedicated to the idea that animals should live as well as people."

So was Charlie Schroeder. Granted, Charlie's earliest transformation efforts were still sterile-looking to today's standards, hygiene being as fanatically important to him as it was, but the improvements were drastic for the animals' comfort and space at the time.

For instance, sleeping quarters—places large enough to accommodate the animals away from the public eye—was first tried in the Children's Zoo. Of all places in the Zoo, the Children's Zoo animals needed to get away from the children every now and then, even though the contact was supervised.

Dr. Schroeder wanted children to be able to touch and play with the baby gorillas, which caused all sorts of hand-wringing problem-solving for Faust. His solution was to build a sun room for the baby gorillas so they could get away from the kids. "They could climb up their little tree and run to the sun room and sit, which gave them a certain security blanket-type of thing," explained Faust. The idea worked well, so Chuck again began refining the concept for the rest of the Zoo.

"One of the world's largest zoos lets animals live like kings," announced a 1959 *Friends* magazine article: "A small tree added to a cage gives a bird new zest for life. Buffalo and deer, which need long pastures to graze in, are given them here. Hippopotami and tapirs are furnished pools of water in which to wallow and cool off. Birds delight in cages large enough to give them fairly long flights."

Charlie had to feel good about what was happening. But the biggest change was next.

CHAPTER 9

A Moat Revolution

"I hate wire," Charlie would say whenever prompted on the matter. For a man who had spent much of his life working with animals in cages, that was quite a statement. Of course, once again, Charlie was ahead of his time. Within a few years, most people, if asked to name one thing about zoos in general that they did not like, would answer that they disliked seeing the captive animals behind bars and wires.

It was time for a big change; Charlie Schroeder knew it intuitively and early.

The innovation in moat enclosures during Charlie Schroeder's director years became the Zoo's defining point in many ways, creating its reputation, and as many of his colleagues testify, influenced a whole generation of zoo transformation across the country.

But the first moat was not at the San Diego Zoo, even though Dr. Wegeforth was the first in the United States to experiment with moats. Dr. Schroeder recalled that the use of grottos and moats was copied from the Hagenbeck Zoo in Hamburg, Germany, one of the world's oldest and most famous zoos. Dr. Wegeforth had visited them in his travels and came back excited at the "free sight enclosures" dating back to the late 1800s.

Also, the Bronx Zoo built a moated Asian Plains exhibit in the forties, and the Brookfield Zoo and St. Louis Zoo had also tried moats experimentally, but Dr. Wegeforth was the one brave enough to try the concept in the Zoo's earliest days.

Charlie must have remembered the sensation those open enclosures caused when he was the Zoo's veterinarian. Imagine San Diegans during the twenties in their bustle and waistcoats and shortpants, rounding the corner and finding themselves being eyed by a lion with seemingly nothing between them but air.

By the fifties, the concept was waiting to be refined, and San Diego's potent combination of excellent year-round weather, a zoo-loving population, and Schroeder's can-do spirit made San Diego the right place.

"It's peculiar how little restraint you need to keep an animal in," he noted. "You can show the animals very effectively with nothing between you and the animal—that's the way it should be. We proceeded to do it. The whole deer mesa was wire, for instance—square pens with cobblestone and hay racks. It was just awful, so we ripped it all out and put in moats."

Soon the San Diego Zoo's moats would be copied by zoos worldwide. In zoo jargon of the time, San Diego was about to move from a "menagerie" into a modern zoo, from "postage stamp" enclosures of one or two animals to a tiny feel of the wild.

If an animal's captive habitat could feel somewhat like "home," then why not do it? If every chain-link fence could be torn down between the animals and the San Diego Zoo visitors, it would be.

"Moat will be the word throughout the Zoo," wrote one reporter describing the transformation happening in Balboa Park, noting that the "Zoo's masters of moatedness are Robert E. Jarboe, superintendent of buildings and grounds, and Zoo designer Charles Faust."

Faust was coming up with plans left and right. Charlie was also bringing back ideas from his zoo travels. He once told of admiring and then copying an exhibit he saw in Holland's Rotterdam Zoo. Later, he saw the director of the Rotterdam Zoo at a conference. As

Charlie explained it: "I said to him, 'I think you should know that we've copied you in San Diego, and it's worked out great.'

"Know what he said? 'Well, we've put up wire.'" Charlie shook his head. "He had a few problems and gave up a good idea."

Wire Comes Tumbling Down

But in sunny San Diego, the wire kept tumbling and tumbling and tumbling down.

From the beginning, Chuck Faust had worked on plans to fix the four existing moated exhibits from Dr. Wegeforth's day. For instance, the bear moats were originally WPA structures built in the Depres-sion. The moats were too wide, too deep, too steep. If animals fell in, it was "some trick" to get them out, as Faust put it. He redesigned them, making the moat V-shaped instead of square and designing the moat with a way to walk out one end. They also experimented with having the animal side lower than the people side, which would keep the animal from having enough space for a long running jump—all based on the observation that an animal cannot make a short run down hill and leap up.

Faust and his cohorts weren't shy about trying things, even if they had to tear it out and try again. For instance, the hoofstock area at first had severe moats, Chuck explained.

"We tried to figure out the safe distance to work with these animals, their flight distance, what would spook them. We had steel bars set inside the moat walls to keep them from getting too close and found out we didn't need any of that, so we took them all out."

Chuck also was doing some creative thinking with glass additions to enclosures, angling the glass, sloping it for best viewing effect all day long. Those ideas are still influencing today's use of glass in the Zoo. He would think of a design, then hire a contractor and "get out there and stomp around the field and decide what we were going to do, and do it." He worked closely with the entire staff of the Zoo, including curators, research hospital scientists, and veterinarians to find out what the animals needed. "It was all a big working family thing," said Faust.

Hours and hours were given to watching the animals. Dr. Schroeder must have found all the animal behavior study fascinating since his zoo veterinarian's curiosity was no doubt deeply piqued. "They know that an alligator can't climb a moat because his chin gets in the way, but that it would take a thirty-foot moat to contain a leopard, which, of course, was space we don't have," Charlie once explained, his fascination apparent. "They found that most animals might accidentally get backed into a moat, but they wouldn't go into a moat by choice. They had a bit of trouble with some belligerent animals fighting and crowding each other into moats, so they moved the bad animals and redesigned the moat until it worked for all the animals in each enclosure."

"When you're planning moats," Faust explained, "you study animals for days to try to decide how far they can jump, then you make your best decision and build the moat. But you are the guy who has to stand there on the first day and see if it works."

That danger is probably the most prevailing reason why zoos such as the one in Rotterdam often reverted to wire, but not San Diego Zoo's wire-hating Charlie Schroeder.

"We think it is possible to put almost any animal in an enclosure without wire," Charlie adamantly believed. Everybody was learning fast that this was a man who could redefine the word "stubborn" when he believed deeply in something.

Impossible took a little longer for Charlie Schroeder, which was exactly how he once described Dr. Wegeforth's attitude. To Charlie, it wasn't impossible to moat monkeys; Charlie's crack "moat" team just hadn't figured out how to do it yet.

But you certainly can't moat monkeys, the keepers told him. They'll get out of anything but wire. They'll crawl right out of any moat devised.

"Then you can put them right back in!" was Charlie's response. Which is exactly what they had to do on several occasions.

When they tried creating an open-air enclosure for the gibbons by adding water to the moat, Chuck Faust and his team slowly pulled down the wire and stood there waiting to see what happened.

The gibbons rushed through the water and right toward where the men stood.

Monkeys 1, moat designers 0. Back to the drawing board.

But that didn't slow Charlie—at least publicly. Moat brainstorming will go on, the visitors will be safe, and the animals will remain contained, he'd say. "Besides, who wants to get out? They're much too happy to leave of their own accord," he once said in widely-quoted remark for the papers.

How much did Charlie hate wire?

Ron Gordon Garrison, the Zoo's longtime photographer, can tell you. Dr. Schroeder disliked wire so much that he would not even allow a picture of a wire-enclosed exhibit to be published. "That was not just a goal he had, it was an absolute demand," said Ron. "During those first years I was here, I was out taking pictures in the primate cages, which back then were just basically sterile, concrete enclosures with a shelf, and maybe a branch or two. The particular day there was a ruffed lemur hanging by its feet from the wire overhead, like a bat would do. It was an interesting picture, and for some reason, the picture wasn't run through the committee before being published in *ZOONOOZ,* the Society's magazine. As soon as he saw it, Dr. Schroeder called me. 'You can see wire in this picture,' he said, then added, 'If you ever do that again, you're fired.' To this day, even though I know we can erase anything digitally, I find it difficult to take a single shot that shows wire."

Nightly Strolls

Charlie's love affair with the concept of a wireless zoo continued the innovative experimentation. It also kept the night life continuously interesting on the Zoo's grounds, because even if the animals did get out, they often came home in the morning.

Years before the Wild Animal Park's concept of grouping animals by the way they lived in the wild, the Zoo experimented with it. Charlie loved to tell how zebras, ostriches, impalas, Thompson's gazelles, and crown cranes were all living in the same large enclosure and doing fine—or so everyone thought. Then he'd add, "We discovered that every night the crown cranes were jumping out of the pen, walking up and down the deer mesa and, before the keeper got there in the morning, jumping back in."

Meanwhile in the Children's Zoo, the unfettered seals were making themselves right at home. The seal pool was designed with a little island in the middle of it, and no restraints anywhere. The seals will stay there, Charlie told Bennie Kirkbride, the Zoo's veteran sea lion show trainer. "But every night," Kirkbride said, they would crawl out and scoot the long way over to sleep in the exit exhibit tunnel, out of the wind, and every morning the Children's Zoo director would have to herd them back to their pool." This went on for as long as the Children's Zoo had a seal pool, amazingly with few real problems. The seals seemed quite happy with the arrangement.

For years, though, Charlie continued to lose the escape-risk debate over the great apes. One day he said to his moat expert, "Chuck, can't you at least somehow get rid of the cyclone fencing on their cages and put up something less distracting?"

Chuck Faust then designed an ape pen with slanting walls to alleviate the vertical planes and lessen the look of wire.

But, of course, Charlie did not give up on moat enclosures for the apes, and today, a visit to the Wild Animal Park's gorilla enclosure is a delight of wide-open space with no wire in sight.

"Nobody had successfully moated gorillas before, so it was a matter of guess work as to how far they could jump, or if they would jump, and what size moat they needed," Chuck remembered. "I can remember the day we let the gorillas out in their moated exhibit. Everybody kind of stood there with crossed fingers, hoping everything would work the way we planned it. And it did."

Lack of Barriers

"The miracle of the San Diego Zoo," wrote columnist Neil Morgan in 1969, "is the lack of barriers between animal and human. But it has not been achieved simply. Charles Faust, who is responsible for much of the recent splendid Zoo design, reminisced the other day about how much has been learned here through trial and error."

And it was true.

"We stumbled on a lot of things," Chuck explained. One time they were going to redesign the giraffe's unit. The giraffes were behind an eighteen-foot chain-link fence, which wasn't exactly aesthetically pleasing for the giraffes or the exhibit. "We decided we would try and moat them, but nobody had really successfully moated giraffes before."

Chuck and Robert Jarboe, the other "master of moatedness," took out a whole section of the fence and put in some experimental moats to see if the giraffes would cross the moat. "The first thing we did was build a four-foot cement wall across the opening and back-filled it on the exhibit side. Then we were going to experiment on what kind of moat to put on the public side. All of a sudden, it dawned on us that these giraffes wouldn't come anywhere near the moat," he remembered. "You could rake the ground in the morning and none of the day's footprints were within four or five feet of the edge of this thing." Like in nature, the giraffes would not step down. It felt too awkward and vulnerable. "We even tried luring them with food," he added. But they never came close, so the giraffes' moat ended up being a very simple containment device.

That fact still amazes Faust: "This small moat, really just minuscule for that large an animal, has worked for years."

And It Works!

By 1959, Charlie was elected president of AAZPA (American Association of Zoos, Parks, and Aquariums), and already he was gaining a reputation for his big-hearted, animals-first way of sharing successes with his zoological colleagues. That made him famous for a few Schroederisms, too. "He was a bundle of energy, very much respected on the national and international level," said Clayton Freiheit, Denver Zoo director. "One of his sayings, was 'It's great . . .and it works!'" An industry magazine's profile about him was even titled that because he was so known for uttering that phrase while enthusiastically sharing his latest wonderful idea.

Clayton tells of such a moment during the first time he ever walked through the San Diego Zoo with Charlie. "They'd just poured a cement walk and used rock salt to create an uneven surface so it wouldn't be slick for people to walk on when it was wet. He pointed down to it and said his famous words: 'It's great . . . and it works!'" Freiheit recalled. "Charlie's attention to detail was legendary. He had the most wonderful ability to be excited about even the smallest things."

He had the same capacity for the big things as well. If something worked, it was worth passing on. His zoo world contemporaries learned it was smart to listen and to come see for themselves.

"We had other zoo directors visit our Zoo all the time," said Chuck Faust, "and we were glad to have them. It's a close-knit family of zoos all over the United States. Anything that works in one zoo, eventually spreads around to others. There are no secrets in zoo design." Charlie told everybody he knew in the zoo world about Faust's sudden insight with the Zoo's giraffe enclosure—because "it worked." In a short time, zoos across the world were following the San Diego's lead and began pulling down their giraffe enclosure's awkward, ugly twenty-foot fences.

Small Drawback

Of course, the moated, open-air freedom had its drawbacks, especially with the more expressive creatures like the gorillas. They had a special way of showing their irritation and opinion of veterinarians who poked and prodded them. Werner Heuschle, director of the Center for Reproduction of Endangered Species, told about showing some German colleagues around the Zoo during his earlier years as a Zoo vet. They were speaking both in German and in English, and as they passed the gorilla enclosure, one of the gorillas picked up a handful of feces and wound up for the throw.

Werner, spying the wind-up, yelled, "Duck, duck!"

And the German answered: "Where's the duck?"

Too late.

Buck Rogers Today

By 1965, Dr. Schroeder proudly told a *San Diego Union* reporter, "There's nothing left of the original Zoo. We have spent more money from earned income since 1954 than Los Angeles has to build its whole great Los Angeles World Zoo. We'll never be finished building. Our imaginations will lead to replacement and new structures forever."

The reporter was then given an impressive taste of Charlie's barrage of ideas. This is the way the article ended: "The talk turns to sound barriers and wind barriers and electricity and other means of containing animals that might sound as way-out today as Buck Rogers did just a few years back. When a better enclosure is built, San Diego's Zoo is sure to have it."

Most zoo directors would pat themselves on the back at this point and turn to other matters, but the revolution in zoo housing and display through moats was an idea that Charlie never let go. In fact, the whole time he was watching and guiding the transformation of Balboa Park's 110 acres, he would be refining an idea that would ultimately give birth to the wide-open 1,800-acre San Diego Wild Animal Park, where boundaries seem more about keeping the people contained than the animals.

"The enclosures were not meant to protect the people; they were meant to protect the animals," Charlie was known to say. By the time Charlie retired and the Wild Animal Park was open, that quote would take on the sound of prophecy.

CHAPTER 10

Making the Zoo "World-Famous"

In 1960, perhaps due to the Zoo's moat success, Dr. Schroeder was asked to write the entry "zoological gardens" for *Reinhold Encyclopedia of the Biological Sciences.*

He began his entry conventionally enough:

> ZOOLOGICAL GARDENS
>
> Living museums, organized zoos, identified as zoological gardens, zoological parks, menageries and aquariums, have been in operation for 150 years. Incorporated nonprofit organizations associated with zoos and identified as zoological. Societies have made zoo operations a scientific program. Notable: Antwerp, Chicago, London, New York, Philadelphia and San Diego.

Then he added this:

HOUSING

> Revolutionary changes are taking place in the design of structures for the exhibition of wild animals. Every effort is being made to permit visitors an unobstructed view of animals leading to the elimination of bars and wire; to bring the specimens up forward for close observations; to associate the animal with plantings typical of its native habitat; to use structural material impervious to urine and feces; to minimize unpleasant odors by the use of ceramic tiles, structural glasses, ceramic coated metals, stainless steels and anodized aluminum.
>
> Plastic sheets and plastic coatings are being used extensively and successfully. It has been shown that radiant floor heat, by means of warm gas, hot water, or other liquids, plastic or lead-covered electric cable, or electric-heated copper tubing, is the ideal, the purpose being to avoid animal heat loss by conduction, convection and radiation. This calls for the use of Styrofoam sheets or aluminum foil built in as vapor barriers to permit floors, walls and ceilings to reflect the temperature of the room or enclosure, and not present cold surfaces. Of equal importance is the avoidance of draft or chimney effect in the holding of exhibit quarters.

Those who knew Dr. Schroeder best must have enjoyed the fact that he was asked to write an encyclopedia entry. As mentioned earlier, his good friend Marlin Perkins once said that Dr. Schroeder had an encyclopedic knowledge of veterinary medicine, but the truth was probably more that he had an encyclopedic knowledge of everything.

"I always wanted to say to Dr. Schroeder, 'How the hell can you know so much about everything?'" Ron Gordon Garrison, the Zoo's longtime photographer, said in sincere awe.

"The speeches he'd give—you never knew what he would say," director of development and former Zoo director Chuck Bieler remembered. "You could give him a subject matter, but depending on what he read last night or the hour before, he might talk about anything. It was fascinating."

"Dr. Schroeder was an expert on everything," commented one longtime employee.

And while having a walking encyclopedia for a boss might have its drawbacks for us lesser mortals, the fact that he never forgot anything meant that a good idea never escaped his memory.

A Way With an Idea

Dr. Schroeder had a wonderful way with an idea, and the San Diego Zoo is what it is today because of his talent for collecting good ideas and his skill at making them fit the Zoo. Perhaps that, coupled with his amazing ability to bring people together, was truly his genius. He was always giving credit for the things accomplished around him, and to his way of thinking, rightly so. Others could have good ideas; he just knew what to do with them. This was a man who could pluck an idea out of the air, and either file it away for the future, take and run with it himself, or hand it off to the right person and, like magic, watch the idea take form and substance.

It was a sight to see. A great example was his strong feelings about flamingos and zoo entrances. Charlie was impressed with the flamingo exhibit at the front of the Hagenbeck Zoo. *What a colorful feast for the zoo visitor's first sight,* he must have thought.

Within months, San Diego had one, too.

As mentioned, when something worked for Charlie, he didn't keep it quiet. "It works!" he said to colleagues near and far.

Within a few years, it seemed an unwritten law that zoos had to "open" with a flamboyance of flamingos, as zoo after zoo added them to the entrance. For instance, flamingo lagoons grace the entrances of the Los Angeles Zoo, the Burbank Zoo, Audubon Zoo, Riverbanks Zoo and Singapore Zoo because Charlie either worked with these zoos' young directors or was hired as the zoos' interim director and installed them himself.

Charlie loved ideas so much that he couldn't let one go, no matter where it originated. Every idea had its chance.

Just a few of the other plans the Zoo would try were escalators that would carry passengers up steep hills in the Zoo, a turtletorium, a hummingbird aviary, a new restaurant with a panoramic view of the grounds, and an aerial tram.

Not all of the ideas were successful, of course. He never quite knew why the tape-recorded Zoo guide for walkers (in English and Spanish), the ten-cent shuttle bus, or the four-person electric cars fell flat on their faces.

He would just shrug and say, "You can never tell about people. You just never can tell."

Looking at Things Differently

Some ideas may have just been ahead of their time. Today, the "Zoophone" which is essentially the same tape-recorded Zoo guide Charlie had in the sixties, is a well-used addition used by "Walkman-trained" visitors.

Another time, he sent a memo to George Pournelle, the mammal curator who was on a trip to Africa, asking him to investigate bringing back a village as an exhibit for the Zoo. It never happened, probably because zoo people still were telling stories about the turn-of-the-century European zoos that actually kept Eskimos and Pygmies in viewing cages for their public.

Still, turn the idea another way and things look very different.

"It seemed absolutely bizarre and demeaning at the time to most of us," said ZOONOOZ editor Marjorie Shaw. "But we have all sorts of cultural fairs and centers now showing other ways of living peopled by representatives of the cultures, don't we?"

Nothing escaped Charlie's active mind; the ideas in place could always be improved or utilized in better ways.

The bus drivers, most of them Zoo old-timers, were already well-known for their ever-changing patter as they drove visitors through the Zoo grounds. Two local columnists, Lew Scarr and Dick Bowman, captured the bus drivers' spiel:

> As the bus passed Roosevelt Junior High, the driver cracked that "over there are the most dangerous animals in captivity."
>
> Other dangerous animals pointed out on the drive include motorists on Highway 163. Down there, you can watch the dangerous animal named the road hog.
>
> The okapi, which is related to the giraffe family, has a twelve-inch tongue and can lick the inside of its ear.

The Zoo is noisiest at night because so many of its creatures sleep all day, especially the big cats.

No two zebras are striped exactly alike.

Green on the polar bears' fur is caused by the algae in their pool.

Hatari the elephant is called that because it appeared in the movie *Hatari!*

The feathers of the rhea once were used for dusters.

Those are red jungle fowl, not banties, running around loose. They are one of the reasons the Zoo has few flies.

Then Scarr added this note to his bus-driver column: "Zoo director Dr. Charles Schroeder has his staff ride the tour bus once a month and take notes."

Who else but Dr. Schroeder would believe that his staff would benefit from taking notes from the Zoo's bus driver observations?

It's Not the Hot Dogs

"A zoo must be a great family center. I have the feeling people like to be with people in crowds," said Dr. Schroeder, who was working mighty hard to create those crowds for San Diego's families and visitors. But he knew people weren't coming to buy the Zoo's hot dogs or to ride its speedy escalators. They were coming to see the animals.

Education was the big reason for the entertainment or it would never be more than a carnival—a sideshow—the one thing Charlie had vowed the Zoo would never become, bowing again to Dr. Harry's spirit.

"A zoo should be more than just looking at elephants," he once told a newspaper. "It should give you a closer look at the fascinating things in nature. When you look at an animal, you should say, 'What a gorgeous creature. Why does it have horns? Why is it that color? Why does it have cloven hooves?' Each animal species has a story to tell."

Then Charlie offered the reporter one of those stories he believed worthy of awe: "Take the emperor penguin, for instance. What a life cycle! How many million years has it evolved to its present life in Antarctica? The male penguin can hold an egg between its feet, even during gale winds. And the males don't eat or drink for three months. They live off their fat and incubate the egg till the wife comes back. And somehow she finds her mate among all the rest. When she returns with fish to feed the babies, Poppa goes off and feeds himself."

The message was this: *Please come and enjoy yourselves, but take something away with you, too, something that will add to your awareness of the world you live in.* As he told a Singapore newspaper during his time there as a zoo consultant: "The zoo plays a major role in our lives, it really does. Kids, especially, should come and see who they share this earth with."

In San Diego, the public came. People were now visiting the Zoo in numbers never imagined, except by Charlie Schroeder, of course.

Building the Collection

"Dr. Schroeder transformed the Zoo with the demolition of the old exhibits," said Jim Dolan, director of collections and former curator during Charlie's era. "That's when the collection took off."

Rare or endangered species became the goal. The Zoo's first okapi came in 1956. Then swiftly came tree kangaroos, langur monkeys, trumpeter swans, and so on. The Zoo's wish list was a long one. The fun was just beginning.

But the fact that there were more and more species becoming rare and endangered meant there was less and less chance to procure animals from the wild and more reason to leave the wild as wild as possible. The fact that the prices were increasing far beyond the norm alerted most zoo directors that something dramatic and serious was happening.

The quick answer was swapping.

Dr. Wegeforth had already paved the way for zoo swaps by creating the forerunner of American Association of Zoos, Parks, and Aquariums (AAZPA) in the twenties, to circumvent commercial animal dealers, who bought up surplus zoo animals and sold them to other zoos at inflated prices. Why buy or sell, decided the enlightened zoo directors, when we can swap with each other?

The association, which now meets annually and publishes a semi-annual "surplus" and "wanted" list, has flourished. By the fifties, swapping finesse was just finding its stride.

Belle Benchley was noteworthy for making landmark swaps with international zoos, so Charlie had her reputation as an outstanding "swapper" to build upon. As the "homey" enclosures created more content animals, they began to breed and the Zoo began to enjoy its first real surpluses. That development, of course, created even more swaps to obtain the rare and endangered species that Dr. Schroeder and his staff truly wanted.

One of Charlie's first big swaps was with, of all places, the Moscow Zoo during the middle of the Cold War. As a goodwill gesture, the San Diego Zoo sent a pair of cougars, a couple of black bears, two raccoons, a crate of rattlesnakes and king snakes, two possums, and a pair of descented skunks. The Russian-bound crates were marked, "From San Diego, Calif. Zoo, United States of America." Charlie and his curators asked in exchange for some Marco Polo sheep, a Siberian lynx, some Siberian tigers, and a couple of Saiga antelope. The entire program of exchanging animals with the Russians took a year of red tape from our State Department and the Russian Embassy.

Boo Boo and Friends

The Convention for International Treaty for Endangered Species (CITES), as well as the United States' Endangered Species Act, did not come into existence until after 1970, so the Zoo was acquiring animals in interesting, eclectic ways, the most interesting coming from sailors returning on both naval and commercial ships who would bring back animals in their holds. One of the last and most interesting sailor gifts to the Zoo was several Gibraltar apes, a species so rare it was called a "relic" species at the time, donated to the Zoo by a British Royal Navy garrison on Gibraltar.

Ironically, as the quarantine regulations began to find their place, sometimes the Zoo would receive animals and birds through them. Such animals would be confiscated, quarantined, and isolated at the Zoo hospital. Then when they'd pass inspection, they'd usually be presented to the Zoo.

One of the most famous gifts, though, was also filled with history and pathos. In November 1965, the U. S. Naval Mobile Construction Battalion was stationed in Chu Lai, Vietnam. Their commander wrote this letter:

November 19, 1965
Director
San Diego Zoo
San Diego, California

Dear Sir:

We have been privileged to have a young Vietnamese Honey Bear (name: Boo Boo) as our mascot. Although we do not know of her exact origin, she is native to this part of Vietnam. She came to us at the young age of two months, by parachute, being dropped in a small cage onto the sandy beaches of Chu Lai.

As our battalion is returning to the States shortly before Christmas, we would like to find a home for Boo Boo where she can be recognized as the mascot of the officers and men of Mobile Construction Battalion Ten, Chu Lai, Vietnam. All the necessary clearance papers for her entry into the United States have been obtained. If it would be possible for Boo Boo to join your great animal family, the officers and men of this Battalion would consider it an honor to present her to you.

Sincerely yours,
T.C. Williams
Commander, CEC, U.S. Navy
Commanding Officer

Charlie's response was a quick one:

December 3, 1965
Commanding Officer
U.S. Naval Mobile Construction Battalion Ten
c/o Fleet Post Office
San Francisco, California

Dear Sir:

Your communication of 19 November 1965 unfortunately did not arrive until Friday, December 3. I hasten to reply in the hope that this will reach you in adequate time.

The college gymnast enjoys the beach with an animal friend

Traveling with Charlie—cross-country tenting took him to vet school

Portrait of a young man on his way to an acclaimed career in the zoo world

Veterinarian Charlie Schroeder in bow tie and big smile with early hospital staff

Charlie Schroeder, D.V.M., treats a camel at the early San Diego Zoo

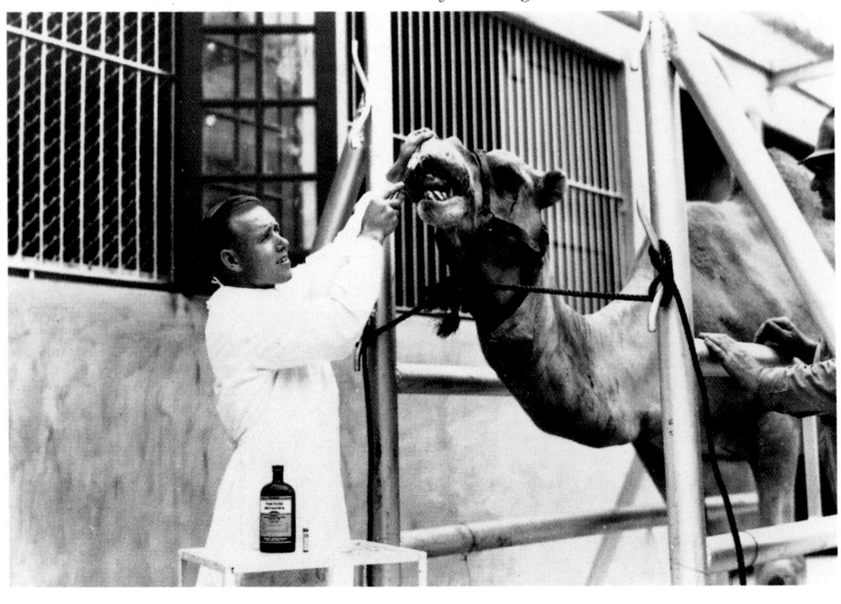

Young Zoo vet Charlie Schroeder on a harvesting trip for sea lions, circa 1935

Dr. Charles R. Schroeder, Zoo vet, in front of the Zoo's hospital building, circa 1939

View of the Zoo's pathology room where Dr. Schroeder performs a necropsy

Zoo vet Charlie Schroeder treats a cassowary with a little help

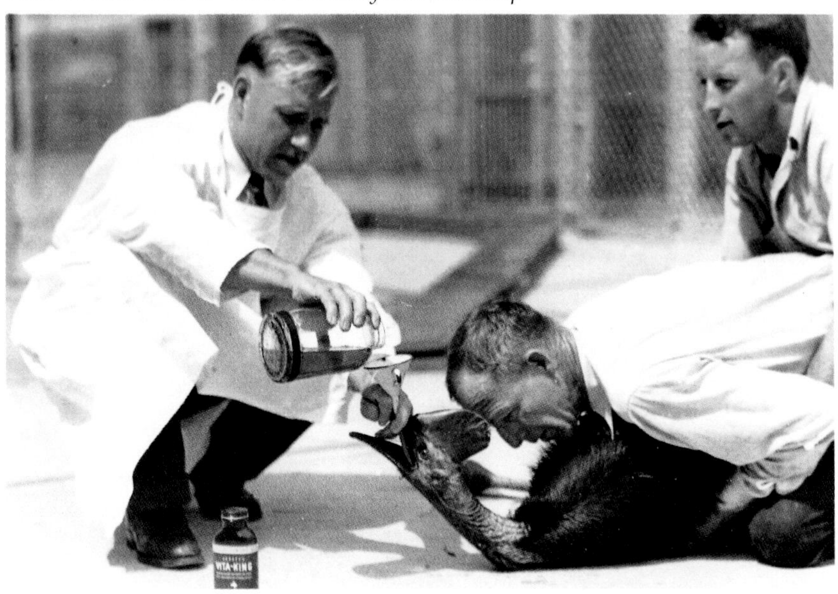

The early San Diego Zoo's entrance

Portrait of Charlie Schroeder, San Diego Zoo Director, 1953-1972

Charlie and bird curator K.C. Lint inspecting the new flamingo enclosure

Today's flamingo enclosure greets Zoo visitors with the same pink brilliance year after year

At the Children's Zoo groundbreaking (from the left), Charlie, Howard Chernoff, Belle Benchley, and Milton Wegeforth with new residents

Charlie with famed anthropologist Jane Goodall and "Daktari" star Marshall Thompson at the Zoo's 50th anniversary celebration

Charlie riding in the Skyfari gondola on its maiden voyage

Sir Charles being knighted in Czechoslovakia, while attending an international zoo conference

Innovative Zoo designer Chuck Faust working on an early moat idea

One of the Zoo's earliest moat enclosures under construction

BEFORE: A wire giraffe cage before the Zoo's moat transformation

AFTER: The new giraffe enclosure vividly shows moat design's advantage

The early gorilla cages of wire and concrete, circa 1950

A lush picture of gorilla life in the Zoo's "Gorilla Tropic" exhibit, 1999

The Zoo's first gorilla moated exhibit, built 1963

Charlie and Margaret Schroeder on safari in Africa

Dr. Schroeder and San Diego Zoological Society honorary life member Arthur Godfrey, at the Children's Zoo 10th Anniversary

In a photo taken by Charlie, Emperor and Empress of Japan visit the hummingbird enclosure, 1975

A photo of Charlie Schroeder, photographer, capturing keeper "Red" Thomas introducing a koala to the Emperor and Empress of Japan

Dr. Schroeder in a photo opportunity with Mouseketeers, keeper Earl Tharsen and cockatoo "King Tut"

Joan Embery and baby gorilla Gordy visit Johnny Carson on the "Tonight Show"

A modern example of the walk-through aviary pioneered during Charlie's directorship, with Skyfari overhead

Several of today's Zoo residents:
Tiger River exhibit's Sumatran tigers
Giant pandas snacking on zoo-grown bamboo
River hippo in Hippo Beach
Polar bears cavorting in Polar Bear Plunge

We will be delighted to receive Boo Boo, your mascot as a gift to the San Diego Zoo.

The San Diego Zoo is U.S. Navy!

Sincerely,
SAN DIEGO ZOOLOGICAL GARDEN
C.R. Schroeder, D.V.M.
Director

What followed was an honorable discharge for Boo Boo, many morale-building ceremonies, heart-tugging headlines, and a children's book, followed by a much-loved life in the Children's Zoo for a Vietnamese bear.

Busman's Holiday

Zoo directors were notorious for taking "busman's holidays" by visiting other zoos on their vacations. Many swaps took place in such settings. A *Saturday Evening Post* profile on the practice told of a Honolulu zookeeper spending part of his vacation at the San Diego Zoo. He was being shown the Zoo by mammal curator George Pournelle, who happened to mention the Zoo had an extra hippo. The Hawaii zookeeper quickly said he'd love to have it but didn't have the money. Right there, in front of the hippo enclosure, the two began talking swap. They found bird curator K. C. Lint, made a deal, then walked into Dr. Schroeder's office for the last word. The last word was "Done."

Dr. Schroeder loved swapping. "Our animal population explosion is due largely to the Zoo's long-standing policy of trading with zoos around the world," he stated in a news story about all the new animals. For instance, in one year, the Sydney Zoo traded twenty female squirrel monkeys and a giant anteater for two tree kangaroos, two white wallabies, two Pesquiet's parrots, one female golden shoulder parakeet, and two double fig parrots.

The Calcutta Zoo offered rare lesser pandas for South American llamas and polar bears.

The Cologne Zoo gave African cape dogs for American bush dogs.

The Ceylon Zoo swapped rare purple-faced langurs for three ocelots and three woolly monkeys.

Of course, the press loved each new arrival:

"THEY'RE HERE! THE LANGURS!" shouted one headline in 1966. "Did you know that there are two purple-faced langurs living in San Diego? The only ones in the country? It's a white-whiskered, leaf-eating, long-tailed monkey that inhabits northern Ceylon."

At some point, one might suspect that part of the thrill for Dr. Schroeder may have been the degree of difficulty; if not, certainly the degree of rarity. The Russian swap wasn't enough. He wanted pandas and that meant trading with Communist China in the middle of the Cold War. "Want to make a nice guy deliriously happy?" wrote local columnist Herbert Lockwood: "All you have to do is deliver a healthy greater panda to San Diego Zoo director Charles Schroeder. He had one coming—it got as far as West Berlin. There, horrified State Department officials stepped in. Please, could he have just one measly panda?"

Ultimately, the State Department gave this interestingly convoluted solution: If a male panda and a female panda could be shipped to a neutral nation, such as India, and produce an offspring there, then the baby panda could receive a visa to enter the United States—but not the mother or father.

Of course, they had no idea what they were asking, since panda breeding is a delicate and rare happenstance. Charlie did not get his panda.

But swapping would become in many ways a peace pipe between the world's zoos, where once there was only competition. By 1972, in a *New York Times* article entitled "Ecology and Conservation Reshape Zoos' Ideas on Animal Care," Los Angeles Zoo director Chester Hogan was quoted about a "mutually beneficial" trade with the San Diego Zoo. They were sending a female Przewalski's horse to San Diego, which only had a breeding stallion, and in return would be receiving a female black rhino that they planned to mate with their male black rhino. "It would have cost each of us several thousand dollars to get them on the open market," said Hogan. What was left unsaid was the fact that beyond the needs of the zoos, these two species were in need of help *from* the zoos—since both were highly endangered.

Another zoo professional quoted in the article put the situation well, saying the "trend to conservation, preservation of species, and the need for breeding have pulled us together."

Thirty-Year Wait

By the late fifties, Charlie had called what was happening with the Zoo's collection an "explosion." That was the correct word for it. These rare creatures were becoming so happy in their moated enclosures, that they began doing what Dr. Schroeder and his curators had hoped: producing offspring. This was where Dr. Schroeder's heart was, of course, and where he already believed the future lay—in captive breeding and research.

For instance, the flamingos began to breed. The flamingos had been at the Zoo since its inception, but not until 1956 did the birds hatch an egg. The difference was their habitat. K. C. Lint brought in mud to add to Dr. Schroeder's new lagoon after investigating breeding success in San Antonio's Zoo and deciding mud was the secret. San Diego joined only three zoos in world that had bred and raised baby flamingos.

The first pygmy chimp in captivity in the United States was born during those years, sired by the same famous pygmy chimp who knew Dr. Schroeder on sight—Kakowet. In fact, the three chimps were the only pygmy chimps in any zoo in the U.S. At that time, fewer than ten lived in captivity in the world, and in their native wilds of the Congo they were becoming rarer and rarer.

Then there was the momentous Galapagos tortoise hatching. Jerry Staedeli, assistant curator of reptiles, recalled a happy moment back in 1958—October 21, to be exact. He had news for the director and rushed into his office to tell him. Dr. Schroeder looked up from his work in a way that stopped Jerry flat. Instead of blurting the news, he asked, "Having a hard day?"

Dr. Schroeder answered that the day hadn't gone too badly.

Then Jerry told him the news he knew was going to make his day go much better. After a thirty-year wait, one of the Zoo's Galapagos tortoises had finally produced hatchlings—five of them. "Dr. Schroeder's smile was a beautiful thing to behold," Jerry said. For years, San Diego Zoo's group was the only breeding colony in captivity.

How Flamingos Lost Their Color

The flamingo story was not only a success story, it also came with a built-in mystery. Right after a new "flamboyance" of flamingos arrived at the San Diego Zoo and joined the current flamingo residents occupying the new lagoon, their pink feathers turned white. What could have happened? Their Zoo home was a lagoon with wading ponds, surrounded by palms, reeds, sand dunes, and grass swales.

The flamingos were safe, seemingly content, and had a wonderful habitat, yet they were fading. Dr. Schroeder called a biochemist from nearby Scripps Institute, Dr. Dennis Fox, to help solve the mystery. He came up with a scientific flamingo diet consisting of rice kibble cooked twice daily, dried flies imported from Mexico, dried shrimp from Florida, dried bell peppers, fresh red beets, plus ground-up lobster shells. That did the trick. The flamingos turned pink once again. When K. C. Lint's mud was placed in their fancy lagoon, the flamingos not only were "in the pink" again, they were breeding for the first time.

The flamingo breeding story was only one of an incredible amount of stories about rare bird species breeding at the Zoo at that time. No other zoo in the world had a comparable record.

At a Break-Neck Clip

"Wouldn't this be a great town if everything were as progressive as the San Diego Zoo?" wrote a *San Diego Union* columnist in 1967. "A livestock inventory there just established that it has the greatest collection of birds ever assembled anywhere: 3,075 birds of 1,008 species. 'We had a nesting season that has Dr. Schroeder all a-twitter,'" said public relations man Bill Seaton.

Pigmy hippopotami, okapi, golden marmosets, Tasmanian devils, Hawaiian geese, Cretan wild goats, Celebes oxen—hundreds of such success stories followed, and Dr. Schroeder and his curators would begin to get a taste for the potential their unique Zoo offered for saving endangered species whose number was multiplying at a break-neck clip.

Something should be done; something could be done. Zoos could be more; zoos had to be more. A difference could be made, right here, beginning now.

Conservation, captive breeding—these words were taking on a meaning that would soon become the rallying cry for the San Diego Zoo with Charlie leading the charge.

By the Zoo's 50th anniversary celebration in 1966, years before adding the Wild Animal Park to its count, the Zoo would be maintaining the world's largest collection of wild animals.

Charlie, The Showman

Meanwhile, with the collection on its way, captive breeding beginning nicely, and research sure to follow, Dr. Schroeder could begin to explore his P.T. Barnum-like ideas.

And that's exactly what he did.

One day, Dr. Schroeder changed the name of the San Diego Zoo.

Meetings and focus groups and votes and more meetings would normally be the steps taken for such a significant undertaking, but that's not for people like Charles Schroeder.

As Jim Dolan put it, "Dr. Schroeder never thought in small terms. One day he just decided to call it the World-Famous San Diego Zoo. That must have driven his colleagues nuts, especially since some of these older institutions have been around since the turn-of the-century or before like Philadelphia and New York. But what happened? He sold it to the world, didn't he? It wasn't that he lied to himself; he told himself the truth and he made it the truth. He made everybody believe it, too. He was very, very good at that."

Public perception is a potent thing; Charlie innately sensed that. The name change wasn't really a formal name change, but more an oral change, word of mouth—yet there is no more potent power than word of mouth. The bus drivers began to welcome their riders to the "World-Famous San Diego Zoo." The Zoo's small amount of advertising began to use that phrase. Charlie began introducing himself as the director of the World-Famous San Diego Zoo.

Within a handful of years, it became second nature to call the Zoo by its Schroeder-decreed name: "The World-Famous San Diego Zoo."

The showman was definitely beginning to show. But perhaps in Charlie's mind, one of the reasons behind this big idea was that it wasn't too far a leap between "World-Famous" and "World's Greatest," which was a goal he would not be denied. His philosophy of blending good business practices with good, innovative zoological thinking was having a revolutionary impact on the Zoo that other zoo professionals were quickly noticing. As famed television host Marlin Perkins would say, "Thanks largely to Dr. Schroeder, the San Diego Zoo has become the richest zoo in the world. He was responsible for upgrading it on a basis so it would support itself."

"What he did was basically set a standard that everyone had to live with. I think it was his bullheadedness," remarked Riverbanks Zoo director Palmer Krantz. "He decided at some point he wanted to have the world's greatest zoo and by God, he did it."

Charlie would not have argued the definition of "greatest" with any detractors, but if greatest meant biggest collection, then by the mid-1960s, there was no argument to be made. The Zoo had the largest animal collection in the world. But "greatest" meant more than that to Charlie Schroeder. It wasn't a competition for him as much as a quest for excellence and trailblazing impact.

"Is the San Diego Zoo the best?" That was the question posed to him later in life by a Singapore newspaper.

"How do you measure?" answered Charlie. "On the basis of operating budgets? Rarity of exhibits or size? No, the criterion is good operations. The BBC made a television series on the eight best zoos in the world. The two in the U.S. were the San Diego Zoo and the Arizona-Sonora Desert Museum. You also need a good director and, like any bank or industry, a good Board. Are they outgoing? Are they going to spend money to make money? Or are they going to sit in a corner, pull in the reins, and say they can't gamble? This whole thing is a gamble."

By 1967, the highest compliment that Charlie's zeal, leadership, and gambling prowess could receive was paid to him by the writer of a *California Veterinarian* magazine profile who made this comment: "When questioned by his interviewer, Dr. Schroeder understandably lost no opportunity to praise the Zoo he directs but which he always refers to as 'our Zoo.' The 'our' is not used in a limited, local sense, although the Zoo is a source of immense civic pride to all San Diegans. Actually, the San Diego Zoo is 'our' to the whole world. . . ."

Whatever the context, by the time Charlie retired, the "World-Famous" had became the "World's Greatest" in popular perception, often to the chagrin of his colleagues he had originally driven "nuts," as Dolan remarked.

During the 1984 Summer Olympics in Los Angeles, the San Diego Zoo purchased five prominent billboards around Los Angeles advertising the San Diego Zoo as the "World's Greatest Zoo." The *Los Angeles Times* recounted the reaction in a long 1985 article: "I don't take umbrage at them for advertising in our market," the article quoted ex-Los Angeles Zoo director Dr. Warren Thomas, "but can't they at least say, 'One of the World's Greatest Zoos'?"

The quoted response from the Zoo's public relations director, Carol Towne, was this: "I've got an inch-thick file of newspaper clippings which say we are the world's greatest."

The *Los Angeles Times* reporter summed it up this way: "San Diego's success in selling itself is perhaps told best by a curator at the Los Angeles Zoo who said with resignation, 'When I go to parties up here and tell people I work at the zoo, they say, 'Oh, the San Diego Zoo?'"

Of course, grasping such a goal took more than bullheadedness. There is nothing like a gifted salesman who believes in his product. Charlie was probably the Zoo's best advertiser all by himself, and this innate gift for marketing seemed to give Charlie pure joy as he put this talent into high gear.

CHAPTER 11

Charlie Schroeder, Savvy Marketer

Andy Phillips can't remember how many times he saw Charlie hand out his special brand of greeting. "Charlie must have gone through thousands of business cards," said Phillips. "He passed them out everywhere. He'd shake people's hands and say, 'Hi, I'm Charlie Schroeder, director of the World-Famous San Diego Zoo. Here's my business card. If you're ever in San Diego. . . .' " To meet Charlie Schroeder was to experience a walking, talking advertisement for the Zoo.

"He showed flashes of genius in many ways, but his touch for the common man was what allowed him to move the Zoological Society forward so fast and so well," said Bill Toone, the Zoo's applied conservation director. "Perhaps that was his greatest strength in everything he did."

A local journalist who interviewed Charlie captured the feeling: "Don't ever tell Dr. Charles Robbins Schroeder you like his Zoo. Dr. Schroeder will flash a wide smile and correct you: 'Our Zoo.' "

It was an attitude that won over just about everyone. Why? The answer has to be his mixture of sincerity and con-man charm. Yes, he wanted what he wanted, but what he wanted was something we all wanted—Our Zoo: the World-Famous and the World's Greatest. How could we not go along with this man's plans to make us part of that?

He was known for calling anyone any time: "I'm Charlie Schroeder over at the San Diego Zoo. I was just thinking about you and wanted to call and say hello." Then most of the time, he'd mention whatever Zoo business he was also just thinking about.

Legend has it that after his first visit to Disneyland, he was so impressed that he picked up the phone and introduced himself to Walt Disney. The result was a longtime friendship. Charlie would send his staff up to Disneyland to learn about cleanliness or interesting design, and Walt Disney would consult with Dr. Schroeder and his staff about the authenticity of his exhibits, from flora to fauna.

The San Diego Zoo was also the first to hire an ad agency, according to Bob Smith, past president of Phillips-Ramsey, the advertising firm Charlie hired. "Charlie would say, 'We are a tourist attraction, but we're going to package ourselves not to entertain but to educate in a very entertaining way.' That was very visionary," said Smith. "He was definitely a trendsetter. He'd say, 'We can have a gift shop, but it must be a great one.'"

For all his charm and self-deprecation, Charlie knew how important the Zoo was to San Diego. Earlier, when Smith was executive director of the San Diego Convention and Visitors Bureau, he remembered a Charlie Schroeder moment.

When it came to tourist attractions, "the world, in that day, revolved around the Zoo—at least the world we knew around here," Smith said. "The Zoo was San Diego. I remember one time—with that twinkle in his eye that always reminded me and my family of Edmund Gwynn, the actor who played Santa Claus in the movie *Miracle of 34th Street*—Charlie said to me, 'Bob, the Convention and Visitors Bureau is great, but, you know, really they ought to pay the Zoo instead of the Zoo paying the Bureau.' But then he'd turn around and be the biggest supporter the Bureau had."

He also never seemed worried about competition for the Zoo. For instance, he believed Sea World to be a wonderful addition to the city, to the point of ultimately offering Sea World the Zoo's penguins. The so-called rivalry remained friendly and mutually beneficial, primarily because of Charlie Schroeder's open-armed attitude.

As for the press, they couldn't seem to get enough of this crackerjack director. Charlie made sure the media was alerted to every happening, no matter how big or how small, because to Charlie all the Zoo happenings were big. Reporters knew they were getting used and they loved it. He may not have been as shameless as Dr. Wegeforth, who'd invite the town to help force-feed their boa constrictor at ten cents a head, dozens paying for the privilege of posing, snake in hand, for the twenties equivalent of a photo op. But Charlie had his ways, and whether he came up with the ideas himself or whether he kept prodding his growing marketing and public relations staff to creative heights, he never let a good news moment pass.

Eggs, Elephants, and Schweitzer

Many were one-of-a-kind stories that no newspaper could resist. One Easter, the Zoo sent an ostrich egg to the White House for its annual Easter Egg Roll with "Happy Easter from the San Diego Zoo," painted on its side. To give exposure to an even worthier cause, the huge egg was delivered by an Easter Seal poster child.

On another occasion, the premier of the Middle Congo announced he was giving President Eisenhower a baby elephant. Upon hearing that, Charlie must have picked up the phone and called the White House and the local press because the next day, a city newspaper headline read: "ZOO REQUESTS IKE'S ELEPHANT." Charlie thought baby "Zimbo" would be a great addition to the new Children's Zoo.

Another time, Dr. Albert Schweitzer's daughter donated a gorilla she had raised at her father's famed African hospital. Its parents had been killed by hunters, and a Peace Corps volunteer brought the gorilla to the hospital. Soon, when the young gorilla began to drop the staff with playful tackles around their ankles, everyone knew he was getting too big to run free, and he couldn't go back into the wild. So Dr. Schweitzer's daughter, after investigating the zoo world for her favorite gorilla, chose San Diego, where he quickly made himself at home.

Editor's New Best Friend

Of course, reporters also relied on Charlie and the Zoo to feed them stories.

After her journalistic encounter with Charlie, journalist Judith Morgan's memory of Dr. Schroeder is of his distinctive voice, a lilt in his walk, and his role in the local news. One of her very first assignments in San Diego was to write a story about the Zoo's latest acquisition—the komodo dragons. She watched Dr. Schroeder introduce the media to the ugliest creatures she ever saw. By sheer stretch of the imagination, she said she was able to write four inches of copy.

Her editor handed it back saying, "You don't understand. This is San Diego, and that's the San Diego Zoo. We've plotted twenty-five inches for the story. Go call Dr. Schroeder because there has to be more."

"By the end of the day, I had my twenty-five inches," she said.

"Charlie was very popular as far as the press was concerned," explained Betty Peach, who covered the Zoo beat during Charlie's tenure. "He was a close friend of whoever was editor. And when that editor left and another came in, within weeks Charlie was his good friend, too. When the Zoo was sponsoring a trip to the zoos of Eastern Europe, he called my editor and said, 'I think it would be splendid if you'd send Betty Peach on this trip. You'd get a good series of stories.' And the editor answered, 'It wouldn't do the Zoo any harm either, would it, Charlie?'"

Betty remembers vividly the West German Zoo, where the touring group saw meandering moats at the same level as the land, animals grouped by their natural habitat, and polar bears with ice in their pool. She also noticed Charlie's high-energy fascination with all he saw. Later, Charlie would be known to bring in ice on ice trucks on extremely hot days to throw into the polar bears' pool and invite reporters for the show.

One of Betty's most vivid memories of being part of the group traveling with Charlie was not writing about Charlie but dining with Charlie. She noticed that after the first night's group dinner, Charlie always maneuvered to sit beside her. It wasn't because he was sweet on her, but sweet on her uneaten sweets—he had noticed she did not eat dessert. "When they came around to ask what he wanted for dessert," remembered Betty, "he'd say he would have the apple pie and my ice cream."

But even if Charlie ate your ice cream, you'd never want to ignore anything he said about animals, Betty Peach found out. All through the trip, he reminded her, "Remember, these are wild animals." He was right. "At the Warsaw Zoo, I was handed a lion cub for a picture," recalled Betty. "We both looked wonderful." But right after the photo was taken, the cub suddenly bit and scratched her. The cub had been hand-reared at the director's daughter's home like one of the family. "But I was not an acquaintance," said Betty. "I found out Charlie was right the hard way. I have scars on my arms to prove it."

Occasionally, the press also published stories Charlie would rather they ignore, such as this item that found its way into print in Neil Morgan's January 14, 1969 column:

> Mon Dieu! Now comes the February issue of *Playboy*, and the dish in the centerpiece is a San Diego girl and that's hardly the most of it. What's most is that among the photographs, there's one of her riding a turtle at the Children's Zoo. When the news reached the Zoo late yesterday, everyone tried to leave before Zoo director Charles Schroeder found out about it. "He'll probably close down the Children's Zoo," moaned one aide.

Teddy Bear, Tiger, or Fox?

"I always saw him as two people—an endearing, laughing, life-loving teddy bear with a man-eating tiger underneath," said Neil Morgan. "Charlie was very good at making you think he wanted your opinion when he was really soaking you with his propaganda, asking, 'What do you think of this?' His sense of strategy was quite amazing."

Another journalist also noticed the savvy side of Dr. Charlie Schroeder and wrote about it:

"The foxy veterinarian who is director of the San Diego Zoo took the press on a special tour of the place yesterday," began journalist Joe Stone's article on a day at the Zoo with Charlie. "He picked a day to show off the Zoo when there was an enormous crowd of visitors," wrote Stone. "He briefed the reporters and photographers on Zoo policy such as the improvement of Dog and Cat Canyon. ('This is a good example of bad exhibition. We will improve it during the next year.') It will look more like the peccary exhibits, he said. ('They are as fine as any you'll see. You can't even smell 'em.')."

During the tour, a fellow reporter asked about what Stone called one of the better Zoo stories of the year: What was being done about the squirrels stealing nuts from the vending machines? "Schroeder said some of the vending machines are fixed so squirrels can't steal from them, but never while he has anything to do with it will all machines be fixed. 'I think it is fun to have the squirrels go in and steal the peanuts. I've seen them do it. It's hilarious.'"

The Million-Dollar Ambassador

Actually, a new member of the media called television became an ambassador itself for the Zoo, making it a household name around the world. In 1954, Bill Fox of KFMB-TV, Channel 8, approached the new director about filming a show at the Zoo, and Charlie knew a golden idea when he heard it.

Beginning on January 9, 1955, *Zoorama* was shot live every week from the Zoo, first with host Doug Oliver then with Bob Dale. It became the longest running program in San Diego history. In 1959, *Zoorama* received the Sylvania Award as "the best locally produced children's series in the United States."

Very quickly, as only television could, *Zoorama* made the World-Famous San Diego Zoo world famous beyond question. In fact, this show's success alone may have inspired Charlie's addition to the Zoo's name. With the invention of videotape, *Zoorama* became a national and then international success, syndicated on some 200 independent stations around the country, and for one year on the CBS network, was the first time a national network selected a locally-produced program for a national slot.

Before it was over, *Zoorama* was translated into several foreign languages, airing in Europe, Japan, Africa—around the free world and back. New York's WOR station offered it five nights a week for a year. Fan mail came from everywhere: Australia, Singapore, Hong Kong, New York, Los Angeles, Phoenix, Honolulu, Ft. Wayne, South

Bend, Kansas City, Albuquerque, Milwaukee, Boise, Bakersfield, Singapore, Sidney, Rotterdam, London. By the time *Zoorama* aired on CBS, it was being seen by sixteen million viewers each week.

Zoorama had become the Zoo's "million-dollar ambassador." Not that it made much money in syndication, but Charlie didn't care. He knew what it was really for—spreading the news of the World-Famous San Diego Zoo around the world.

Bill Seaton, the public relations manager hired to work with *Zoorama,* knew the world-famous truth by personal experience. "Charlie would send me to different international conferences," Bill remembered. "I'd be changing my tickets in London, and the woman behind the counter would say, 'San Diego! They've got a wonderful zoo. We see it all the time on the telly.' Or I'd be sitting in a hotel bar in Africa talking about San Diego and the Zoo and someone would invariably say, 'Oh yes, San Diego—we watch *Zoorama* all the time here.'"

Zoorama was the most remarkable show ever put together "with bailing wire and string," Bill would often say. "In Middle America it was like the Johnny Carson's *Tonight Show*. It really put the Zoo on the map."

A fan letter from Amherst, Massachusetts, echoed the sentiment of most of the letters sent: "We watch your Zoo show every Sunday. I would love to visit your Zoo."

Ralph Hodges, producer and director, attributed the show's success to Schroeder's innovation and drive. "He allowed us to go into areas to show the public what was really happening at the Zoo. He had the courage to tell it all as it is."

"He insisted the animals never be harmed, refusing to allow the common show-business practice of withholding food to encourage a performance," remembered Seaton. "He wanted the show to be scientific and educational as well as entertaining. 'The Zoo is not a circus,' he'd tell us. 'We don't particularly like hats on the animals.'"

The topics of the shows reflected that. "We did shows on interesting things people wouldn't know much about," said Bob Dale. "Like the fact that mice fed to the pythons had to be genetically perfect because the snakes were worth a fortune. Or we'd set up contests to see which was stronger—the gorilla or the orangutan. Or we'd watch to see if Albert the gorilla would use his built-in tom-tom in his cage.

"Sometimes we'd learn things that even the keepers didn't know, such as how a baby hippo feeds underwater," Bob went on. "We had a new underwater camera that had been developed for the hit television series *Sea Hunt*, so we were all excited about using it. The show turned out exciting, all right, but for more reasons than we expected. We were introduced to 'hippo missile mama.' We found out how a hippo moves fast through water. In the middle of our taping her feeding her baby, she decided we were threatening the baby hippo. The film showed what happened next—she gulped air, submerged, and then let the air go which made her come out of the water right at us like a big submerged beach ball would do."

Almost as nerve-wracking as missile hippos, though, were the rare times that Dr. Schroeder was on the show. "Every time he was on the show, it felt like working with God," said Bob. "I didn't want to offend the man or upset him in any way, shape, or form. It was like trying to please a parent or teacher."

The few times Charlie appeared on the show to talk about the newest improvement such as the new escalator or walk-in free flight cage, he refused to take credit for the innovations. "The man was so modest, he wouldn't take credit for anything," Bob recalled. "I'd be sweating bullets to set him up, such as, 'Without you, Dr. Schroeder, we'd still have gorillas in cages.' He'd just give praise to Belle Benchley or Dr. Wegeforth or one of his staff."

No Rain

Charlie was even known to control the weather—at least for television. In the *Zoorama* world, it was *always* a sunny day in San Diego. "Dr. Schroeder didn't want people to know when it was raining, because he didn't want anyone to think it ever rained at the San Diego Zoo. So when it rained, the crew would have to do the show in the reptile house since it was the only place with a roof on it," Bob said. And if they couldn't do that, then Charlie would ask the crew to manufacture sun.

"If the streets were still wet, we'd put on the sun guns which were real bright lights to make it look like the sun was shining," Bob went on. "But we also would have some landscape people in the background with hoses, like they were cleaning the streets. So to the viewer, it never rained here, but, boy, was it clean!"

The biggest tension on the set, though, was the conflict between Dr. Charles R. Schroeder and Charlie Schroeder—the inner tension between his scientific side and his savvy marketer side. Charlie wanted viewers, but Dr. Schroeder also wanted the respect of his Zoo colleagues for the show's content. It was a tightrope that Bob Dale said they were always walking.

"We were making the show colorful and family-oriented, and he kept telling us to make it more scientific," explained Bob. "I played the hick from the sticks, the guy who doesn't know squat about animals and who comes up with crazy questions like 'Are those zebras black with white stripes or white with black stripes?' Dr. Schroeder would warn me sometimes when he thought I was playing it up too much. I'd say, 'But Dr. Schroeder, our audience is just the average guy. The people who watch the show are not zoologists. First you got to get their attention.' Dr. Schroeder, though, was much more interested in the world seeing the scientific aspects of what we were doing.

"We had this discussion for years—back and forth and back and forth. I could always tell when he was irritated, too. He'd have that big smile, but his shoulders would begin to shrug and twitch a little—like James Cagney's, muttering 'All right, you guys!'" Bob remembered, laughing.

"Oddest thing, though," he added. "We did a show week after week after week all those years, and somehow even when Dr. Schroeder wasn't there, he was there. It's a strange thing to feel somebody's presence all the time, whether he was there or not. It certainly made you feel that you'd doggone better do a decent job. He was strong, like an iron fist in a velvet glove. Without him, we just wouldn't have become the World-Famous Zoo."

It must have been a constant battle for Charlie. His Mr. Science sensibility was undeniably strong, but it could still be tempted by a show-biz idea if it were the right one, and Bill Seaton came up with a legendary one.

Baby Elephant Walk

In 1967, to commemorate the tenth anniversary of the Children's Zoo, *Zoorama* gave away an elephant. The Zoo had a surplus of elephants so Seaton thought giving one away on the *Zoorama* show would create terrific publicity. He made the proposal, then stood back for Charlie's response.

To his surprise, Charlie loved it. "He said, 'By God, we could do that,'" Seaton remembered. Of course, he'd also have to defend the concept to the Board of Trustees, but somehow he did, and the contest was a go.

The prize elephant would be awarded to the child who sent in the best birthday card and who happened to live in a city whose zoo was in need of an elephant. Among the judges were longtime Zoo supporter Art Linkletter and La Jolla author Theodore Geisel, better known as Dr. Seuss. On the celebrated day, Arthur Godfrey appeared on *Zoorama* to present the elephant to the winner, a boy from Arizona.

At first, Bill was disappointed the giveaway didn't draw the kind of national publicity he thought it would. But then, suddenly, it was making headlines everywhere. When the elephant arrived at the boy's hometown zoo in Arizona, the Internal Revenue Service demanded that the owner—a boy—pay taxes on his valuable prize. As you might imagine, the press had a field day with that story and within days, the IRS quietly withdrew its request.

By 1970, *Zoorama* had run its course, partly because it was unable to keep pace with high-budget network competitors such as *Wild Kingdom*, but everyone agreed it had been quite a ride.

Miss Zoofari Comes Along

With the end of *Zoorama*, Dr. Schroeder hated the loss of all the free national advertising. At the same time, Board members kept asking Seaton if there was some way to get the Zoo on the *Tonight Show*.

"I said I didn't think so," Bill responded. "Johnny Carson was well-known for not liking animals."

Walt Disney once told Charlie that the Zoo should have an ambassador like Disneyland's, who traveled around the world representing the Disney theme park. She was a pretty young woman, a model whom Charlie had met when she visited the Zoo in her ambassador role. So when Bill Seaton suggested a "Miss Zoofari," Charlie thought it a great idea. Bill interviewed hundreds of young women and came up with a model-type much like

Disneyland's to represent the Zoo. But she didn't last long. After all, she was a model who was being asked to handle animals, and she didn't do very well with them. The decision was made to replace her with someone who was good with animals and could hold her own representing the Zoo in interviews and speeches.

At that time, an eighteen-year-old teen named Joan Embery was doing on-call work with the Children's Zoo while taking college classes. Very few Zoo positions were open to women handling animals back then, and all of them were in the Children's Zoo. She answered an ad that drew 10,000 applicants, so she had jumped at the on-call spot, even though she had to juggle her classes to work whenever they needed her. Soon she was swapping hours with other women who didn't like working with the large animals, especially elephants who were famous for pushing and knocking around their keepers. Joan, though, fell in love with them. She was used to working with horses, and she had a special bond with large animals. When the elephants would push her she'd push right back, telling them to "knock it off." During that time, she began to train a baby elephant named Carol and soon, she was teaching her to paint.

When the Miss Zoofari position opened again, she interviewed. "Most of the applicants were women who thought this would be their big break into television. I applied because it would be better hours and I could have more freedom to work with the animals," Joan remembered. "For me it was a major step because I was shy. I did not like public speaking at all."

During that time, though, she had already caught the eye of Charlie's second wife, Maxine. Charlie was talking about the problem when Maxine said, "Why don't you try Joanie, the Embery girl? She's very good with animals." The next day, Charlie called Bill Seaton and suggested he take a closer look at the young keeper.

She interviewed again. Quietly, after the first Miss Zoofari hoopla, she was offered the position on a trial basis, and that began her long career as the Zoo's goodwill ambassador.

She'll never forget Charlie taking her to her first Rotary meeting. She thought she was there to escort some animals for Charlie's talk.

"Next thing I know, Charlie is introducing me, motioning me to take the podium, and I panic," Joan remembered. "It was like I'd been thrown in the middle of the ocean and didn't know how to swim. To make matters worse, whatever I was going to say would be in front of my boss who knew everything. But you didn't tell Charlie Schroeder, 'No, sorry, I can't do that.' Instead you just somehow do it. And that's what I did. I'd never been in front of cameras or crowds. And I never would have, except I wanted to handle animals, so I always made sure I had an animal with me everywhere I spoke, as a crutch. I knew if I had an animal, the audience would be looking at it and not at me. Charlie had great confidence in me, and his confidence gave me confidence. My career exists because of not wanting to disappoint him." Her way with animals made the difference and soon, Joan was speaking and appearing consistently with her animal companions, especially with Carol, the young painting elephant.

Within a short time, Joan and Carol were doing the impossible—appearing on the *Tonight Show.* In 1971, in her first of more than 100 visits on the *Tonight Show,* she escorted Carol to meet Johnny Carson and to paint for him while millions watched nationwide. Some of the nation's best late-night memories are of Johnny mugging with the San Diego Zoo animal ambassadors that Joan Embery, the Zoo's goodwill ambassador, brought to the show. "It had never been done, live animals on a national television show," she pointed out. Joan, her ever-changing animal companion, and the San Diego Zoo became almost a staple of Johnny Carson's show during those years.

Skyfari High

Another of Charlie's ideas for the Zoo was sky-high and again, unprecedented. On his daily walks, he noticed that a large majority of the visitors never visited a significant portion of the Zoo—the exhibits farthest from the front gate. The Zoological Society had spent $2.5 million on improvements in the rear sections of the Zoo, but very few people were finding their way back there. Charlie thought, *Why not put in an aerial tram that people would enjoy riding so much that they'd pay a buck to transport them back there?* After all, it would offer a whole other view of not just the Zoo but Balboa Park and the city as well. How could it lose?

Soon, he was putting all his formidable Charlie Schroeder persuasion skills to the task of getting the concept off the ground and in the air.

Where did he get the idea for the Skyfari? Disneyland? Who knows. Yet he knew instinctively that it would be successful, giving visitors a new view of the Zoo and bringing in needed revenue. Of course, that didn't mean it wouldn't create hurdles for Charlie to vault. For starters, it had to be technically considered transportation, not an amusement ride, or else it would be taxed.

Charlie went to the Zoological Board and asked for $300,000 for his new "mode of transportation." They balked. Soon, however, Charlie persuaded them to endorse the idea of a cable car "mode of transportation."

That's when things got interesting.

Sheldon Campbell, a future Board president, was once quoted as saying that the tram illustrated Charlie's "mastery of managerial strategy." The story goes that Charlie spent the $300,000 and all he had to show for it was large pylons in the ground. The estimate was far too low. He went back to the Board and asked for more money. How could they say no? Campbell, a stockbroker, always believed Charlie intentionally underestimated the cost in trying to sell the idea to the Board. "The technique was well known. It's called 'buying in.' The Pentagon does it all the time. I suspect he did it deliberately. When he wanted something, nothing would stop him. There was a large element of bulldog in him."

"It cost $2 million to finally build," Board member Eugene Trepte remembered and then added with a wry chuckle: "Charlie seemed to always get his decimal points mixed up."

Whether he left off a zero intentionally or whether he was willing to gamble that his instincts were right no matter what the cost, we may never know. Whatever the truth was, it didn't matter, as Campbell himself pointed out, because Skyfari was a financial success from the very first year it was open—1969. For its grand opening, Art Linkletter, a baby elephant, and a baby gorilla did the honors. Everybody wanted to take this new "mode of transportation" to the back of the Zoo and sometimes, tired and footsore, take it back to the front of the Zoo, too. Skyfari recouped the $2 million and was making that much each year. By April 1970, its millionth rider was honored. The Zoo's rear sections would now be appreciated as they should.

That's the way it is even today. The Skyfari cable car aerial tram rides high over the Zoo.

"Most everything the man did was a first for the zoo community," Andy Phillips points out. "Somehow, Charlie was always so far ahead of everyone else."

Charlie being far ahead of everyone else, though, also seemed to make things happen that no one could expect.

CHAPTER 12

Hollywood, Trees, Vultures, and a Fortune

Sometimes the stories told about Charlie Schroeder during the height of his directorship sound more like fable than fact—tall tales about a larger-than-life guy. How many zoos could get electrical power turned off in a whole section of the city to transport a tree?

In 1959, when the Zoo had a valuable seventy-five-year-old Senegal date palm bequeathed to the Zoo from an estate in Ocean Beach, Charlie somehow persuaded the city to turn off the electricity along the way in order to transport the tree under heavy power lines, even getting some of the power lines moved. According to reports, the Zoo paid a $2,500 moving bill to bring the twenty-six-ton tree to its place of honor on the Balboa Park Zoo grounds.

Jane Goodall, who visited the Zoo for its 50th anniversary, was awestruck at the story. She stared at the palm hovering over her at the Zoo as Dr. Schroeder and Board Member Sheldon Campbell related the tale.

"Surely no one could actually transplant a tree this size. The roots must go down for miles," was her response that day. Years later, she wrote: "That incident made a deep impression on me. This was no ordinary zoo if it could successfully tackle a problem of such magnitude. It was a zoo which up to that time had been highly successful, and a zoo which obviously had a great future."

Charlie valued plants and habitat as much as the animals. As he and his curators worked tirelessly to build the animal collection, he also guided the growth of the Zoo's plant collection to the point that the Zoo became a registered botanical garden. Most of the plants originally were acquired for the creation of habitats authentic to the animals' homes, while the rest were being grown and harvested for animals' strict diets, such as eucalyptus and acacia, which were staples for giraffes, koalas, okapi, and the great apes.

Today, some of the Zoo's plants are as endangered as the animals themselves. The plant collection is so rare and unusual that it is actually worth far more than the animal collection.

All this foresighted concern about the Zoo's habitat was probably just a natural part of the Zoo's growth to Charlie. After all, he had worked for San Diego's own Johnny Appleseed, Dr. Harry Wegeforth. Dr. Harry would take train trips across the country and would always come back with seeds and samplings to plant at the Zoo. He had a cane with a handle on top and a sharp, pointed end. Charlie remembered how he'd poke it in the ground, drop a seedling in, and stomp it with his foot, saying, "If half these things grow, we'll have a great place."

More than half did grow, the lush garden result was so obvious, it soon began to gain notice of its own. A *New Yorker* magazine writer named Emily Hahn wrote glowing words about her San Diego visit in a 1967 book on zoos called *Animal Gardens.* Although she knew the Zoo had changed considerably since her last visit in the thirties, "no written or oral description had quite prepared me for what I saw," she wrote. "I didn't recollect anything of this palatial estate with its heavily wooded surroundings. A big reason for my mystification, I learned, was that there is such a lot of green stuff growing. Dr. Schroeder later told me that the Zoo's botanical collection—which being valued at $2 million is worth twice as much as all the animals put together—contains very few indigenous species."

Zoo's Film Stars

The list of celebrities and dignitaries that trekked through the World-Famous San Diego Zoo during Charlie's years was a long one. Many of the Hollywood stars were Zoo residents, always good for a photo opportunity and few lines of news copy.

From Dinah Shore to the Mouseketeers to Zoo-lover Jimmy Stewart, the San Diego Zoo was always catching Tinseltown's eye. For a while, even before Dr. Schroeder's years, the San Diego Zoo was almost Hollywood South—if not to film a movie, to research one. Johnny Weismuller, for instance, came to the Zoo to study gorilla movements before making his *Tarzan* films. *Rampage, Moon Walk,* and *Hatari* were filmed at the Zoo during Charlie's years. Hatari, the baby elephant who starred in the movie, came to the San Diego Zoo after the 1961 film and lived a long and treasured life there.

The camel featured with Bob Hope, Bing Crosby, and Dorothy Lamour in the 1942 film *Road to Morocco* was a Zoo resident. Sheik, a Zoo camel, also starred in *Beau Geste,* commanding top dollar because he didn't bite.

The Disney Studios photographed the birth of a fawn at the Zoo that was used for the animation of the classic children's film *Bambi.* The prehistoric monsters in the Hal Roach Production of *A Million Years BC* with scantily-clad Raquel Welch were little iguanas from the Zoo filmed inside miniature sets. In 1957, Diane DeMille, granddaughter of the Hollywood film producer Cecil B. DeMille, posed for a news photograph at the San Diego Zoo with the water buffalo that her grandfather had used in *The Ten Commandments* and was now donating.

There was also the occasional dignitary who would pay a visit to the San Diego Zoo. Such a memorable visit happened in 1975 when Emperor and Empress Hirohito of Japan dropped by. The royal couple were interested in the hummingbird aviary, but they did not want photographers hounding them. Chuck Bieler, the director at the time, knew that Dr. Schroeder was handy with a camera, and he also knew the Emperor and Empress would not mind his company. "Charlie was the only one permitted to have a camera as they walked into the aviary. He was able to catch the Empress feeding one of the birds who had landed on her hand," said Chuck. "The photo made the papers coast to coast, but with no byline."

From the Emperor's and Empress' response, Chuck said one of the most prized keepsakes of their trip was the picture by Charlie Schroeder, photojournalist.

An Unexpected Gift

During the sixties, the Zoo reaped an unexpected publicity bonanza that would start some very big dreams on their way, dreams as big as the Wild Animal Park, just because a man named Elmer Otto loved *Zoorama.*

One day, Charlie received a letter from a friend at San Diego Trust and Savings that said, "Dear Charlie, you won't have to worry about feeding the elephants this winter. We're just settling an estate and it's left you $15,000." Well, that didn't sound like a lot to Charlie, but as it turned out, Charlie misread the number of zeros. The amount wasn't $15,000; it was $1.5 *million.*

The interesting thing was no one at the Zoo had ever heard of Elmer Otto. He lived about fifty miles east in Alpine on a modest ranch. Elmer never visited the Zoo, but he watched *Zoorama,* which he obviously liked very much. He was so modest that many people who knew him were surprised to learn he was rich. Otto would even drive into town in an old Buick wearing dungarees.

"When his doctor found out he'd left the zoo a million and a half, he almost fainted because he was one of his charity patients!" Charlie said. "Much of the money was in stocks, and Andy Borthwick [a longtime Board member] down at the bank told us to hold onto it for a while before selling. Well, the stocks doubled, so we had even more money by the time we sold. That's when we decided to build the Otto Center," Charlie explained, "and since we had money left over, we set it aside and would draw from it whenever we needed money. A lot of it went into the Wild Animal Park to get it all started, right when we needed it. Extraordinary, wasn't it? And no one at the Zoo ever met Otto!"

Education at the Otto Center

The unusual, round building on the edge of the Zoo has been the site of learning since 1966. Education had always been an important part of the Zoo, especially since Dr. Harry put the Zoo in trust for the city's children. But now with this money, the Zoo was able to focus on it. The Elmer Otto Memorial Center was completed in October 1966, and since then thousands and thousands of children have been exposed to the earth's wild animals and by extension, the world itself, through its programs.

The ways to educate children and adults alike were now only limited by the new staff's imagination. The new programs ranged from preschool through college level. Instructional tours became a constant part of a day in the Zoo. Thousands of students were admitted free as part of the programs to enjoy a learning experience in this "observational laboratory," as one educator called it. School bus tours through the Zoo were begun in 1926 with a donated fifteen-passenger Model T Ford. Second-grade students from across the city who visited during the year as part of the public school's curriculum were now riding in new buses. During summer, zoology classes for kids were held in the Children's Zoo. Otto's money went a very long way and today, the Otto Center is still being used for a host of educational programs.

Big Bear and Charlie

Lest we forget how much an impact that a zoo can make on a child's life and future, Charlie certainly knew. Long before there was an Otto Center, Charlie was taking time to make a lasting imprint on young Zoo-loving minds.

Andy Phillips remembers vividly the first time he ever saw Charlie Schroeder.

"When I was about six years old, my mother enrolled me in Zoo summer school. The first day of summer school the students received an orientation that included meeting the director," Andy said. "The teacher gathered the group of students together and mentioned that we were going into the next room to be introduced to the man who was in charge of the entire Zoo.

"As we filed in, there was Charlie. Dr. Schroeder was not a tall man, but he was certainly a showman. He knew how to make an impression. In typical Schroeder style, he was standing right in front of an enormous stuffed bear, at least nine feet tall, with its paws directly above his head, which looked all the bigger because of Charlie's short stature," Andy said. "He had to have planned this. I can't remember what he said that day, but I will always remember his grin and that bear."

Andy would grow up to be a comparative physiologist who would spend years in the field, ultimately becoming deputy director of the Zoo's Center for Reproduction of Endangered Species (CRES) and program director for the National Science Foundation's Ecological and Evolutionary Physiology.

A Man of His Word

Bears and Charlie Schroeder were Andy Phillips' introduction to his future. For Bill Toone, it was chickens and a Charlie Schroeder letter.

When Bill Toone was six, his parents took him to the Zoo, and he fell in love with a red chicken that he found underneath some bushes with her baby chicks. During those years, the chickens—Indian red junglefowl—were allowed to run free through the Zoo. Bill's parents were trying to show him the rest of the Zoo, but once he saw those chicks, he was on his hands and knees looking under that bush. The Zoo might as well have vanished.

"I must have driven my parents crazy because by the time they got home, they sent me to my room to write a letter to the bird curator, K. C. Lint, to ask him where I could get one of those chickens," said Bill. "Even then, I had a concept of how important and busy the people were at the Zoo. So when K. C. Lint called three days later and told my mom and dad that if they'd bring me down to the Zoo, he'd give me some chickens, I was floored."

They took Bill to pick up his chickens, and within a year he had a backyard menagerie of pheasants, finches, ducks, and chickens. "I was bird curator of my own zoo," Bill remembered.

So at seven years old, he sat down to write the Zoo again. This time he wrote Dr. Schroeder and asked for a job, and not just any job—he wanted to be the San Diego Zoo's curator of birds.

Charlie read the letter and wrote back a response that changed the course of Bill's life. The year was 1962.

"Dear Bill," it said, "we really appreciate your interest, but you're too young to drive a car, you're too young to join the union, and you're too young for our insurance to cover you. But I have this dream of a North County campus close to where you live, and by the time that dream becomes reality, you'll probably be sixteen years old. You'll be able to drive, and you'll be close to the park. Come see me, and we'll get you a job."

He took Charlie at his word, which turned out to be a good thing to do. "On my sixteenth birthday," said Toone, "I brought his letter to the director of personnel, Tim O'Farrell, and told him I was ready to go to work. The Wild Animal Park was still under construction but true to his word, they gave me a job even then, picking up trash."

By 1980, Bill was in charge of the Zoo's arm of the California Condor Recovery Project. Three years later, he was assistant curator of birds at the San Diego Zoo based at the Wild Animal Park. He had become close friends with Charlie, and at Charlie's death he was asked to speak a few words at memorial service held at the Wild Animal Park. Today he is the director of applied conservation, working around the world to encourage governments to protect endangered habitats, establishing such endeavors as butterfly farms for indigenous populace as an alternative to progress' havoc. He believes strongly that his work is taking Dr. Schroeder's vision into the next generation.

Schroeder the Vulture

Of course, neither Bill nor Andy were immune to Dr. Schroeder's black book memo habit, even after his retirement. As much as Bill enjoyed his talks with Charlie each time he'd see him at the Wild Animal Park, he also knew it would probably not be the last he heard from Charlie that day.

"It got so you knew that if he talked to you, it probably meant you were about to get in trouble. Somebody was going to call you in a minute and tell you to tuck your shirt in or get your hair cut or whatever because, even as director emeritus, Charlie was paying attention to detail," Bill remembered. In fact, Bill and his California Condor team nicknamed one of their vultures "Schroeder."

"He was the one who would be nice and sociable and fun when you were visiting him in his cage, then run around behind you and bite you," Bill said. Charlie knew about his namesake and laughed the hardiest. "Charlie knew he did that," Bill added. "because as much as he liked us, he knew why we were all here. So did we—and that always, always came first."

CHAPTER 13

Beloved Taskmaster

Make no mistake about it. While Charlie might spend $2,500 to move a donated tree and conveniently forget where to put a decimal point on a sky-high idea, Dr. Schroeder, Business Manager, was a frugal, watchful leader. He was frugal with people, he'd say, so he would not have to be with animals.

One well-known story concerning the Zoo's way with a dollar involves a tiger and a lunch box. A Bengal tiger wasn't very happy with his keeper one day. The keeper, lunch box in hand, was standing in front of the exhibit, looking down at the moat before beginning work. When the Bengal saw him, he rushed into the moat, then snarling, jumped straight up from the bottom with his feet, claws, and forelegs up on the moat's square end, as if he were about to come right at the keeper, moat or no moat. The keeper swung his lunch bucket, bopping the tiger over the head, which made the surprised tiger fall back into the moat and rush away. Meanwhile, crisis averted, the keeper saw that his lunch bucket was squashed, so the keeper and his curator supervisor went to Ella Hoover, the Zoo's comptroller, to get some money for a new box. The legendary Ms. Hoover had nothing in the budget to pay for smashed lunch buckets, and she told him so.

"When you wanted a new pad of paper or a new pencil—a pencil, not a box of pencils—at Society expense, you had to go see Ella Hoover to show her your old pencil stub before she would issue you a new one," remembered Andy Phillips. "You can't tell me Charlie didn't control the finances through her."

While he raised salaries drastically from Belle Benchley's era and ultimately managed the Zoo into arguably the most financially sound zoo in the world, the bulk of the Zoo's money was always earmarked for the animals. The Zoo's salaries remained low and to work there was to *want* to work there, which may have been Charlie's point.

Bill Seaton remembered how he finally got his job in public relations. "I was the third pick. Two others had turned it down because of the low pay. 'Sunshine dollars,' we called it back then." Yet even with the low salary, Bill remembered there were 138 applicants for the Zoo job, which was unheard of at that time.

On the Team

Charlie had a team mentality about life at the Zoo, almost as if working there was not a job but a calling. He was known for beginning a meeting by opening his mail and sharing it with the group. He always assumed his employees wanted to know his business since his business was the Zoo. "He irrepressibly liked company and conversation," wrote *ZOONOOZ* editor Marjorie Shaw in a profile about him. Even after retirement, "he hailed employees to his lunch table, always keeping track of the pulse of the place."

Since anyone on the Zoo's "team" had to be as sold out for the Zoo as he was, surely he or she would not mind taking on more and more tasks for the team effort. After all, he expected his people to rise to the occasion and wear as many hats as he gave you to wear. Just ask Chuck Bieler, who was hired to begin a group sales department. One day Chuck got a call from Dr. Schroeder.

"Chuck, I'd like you to be in charge of the buses," he said. "Can you handle that?'

"Am I still in charge of groups sales?" he asked Dr. Schroeder.

"Yes, yes," he answered. "Can you do it?"

"Certainly," said Chuck.

Later, after the Skyfari was up and running, he got another call from Dr. Schroeder.

"Chuck," he said this time, "I'd like you to be in charge of the Skyfari."

"Am I still in charge of group sales and the buses?" he asked Dr. Schroeder.

"Yes, yes. Can you do it?"

"Certainly," said Chuck.

After Charles Shaw, the Zoo's reptile curator and assistant director, died, Chuck got another call from Dr. Schroeder.

"Chuck, I'd like you to make your desk here," he told Chuck.

"Does that mean I'm the assistant director?" Chuck asked Dr. Schroeder.

"No, no. We'll call you my administrative assistant."

"Well, Dr. Schroeder, does that mean I'm no longer in group sales?"

"No, you're still in charge of group sales."

"And still in charge of the buses and Skyfari?"

"Yes, yes. Can you do it?"

"Certainly," said Chuck.

Then one day, while the Wild Animal Park was being built he called Bieler into his office. "Now, Chuck, we need somebody up there to keep an eye on things. We thought you would be a good guy for that," Dr. Schroeder said to him.

"You mean I'll be the director of the Park?" asked Bieler.

"No, no, you just go up there and make sure everything's in order."

"You mean operations director?"

"We'll figure that out later."

"But I'm still director of group sales?"

"Yes."

"Still in charge of the buses and Skyfari?

"Yes."

"Still your administrative assistant?"

"Yes, yes, but you also watch over the Park. Can you do it?'"

"Certainly," answered Chuck.

Thankfully for Chuck Bieler, who would later become Zoo director in both title and responsibility, the list stopped there.

Who's Feeding the Animals?

Dr. Schroeder didn't like the concept of a union, which by definition allowed "outsiders" some of the cherished control of the Zoo. In fact, he detested the idea. Inevitably, though, the employees voted for union membership, and he responded so vehemently according to accounts of employees involved, that it was almost as if he took the vote personally.

But to Charlie, the Zoo was personal.

He loved an incident that happened during the one short wildcat strike called by the union.

On that day, the keepers were picketing in front of the Zoo. "It shocked everyone since the town loved the Zoo," Bill Seaton remembered. "It drew people from across the city to watch, until finally one little old lady strode up to the picketers, pointed inside the Zoo, and yelled, 'I go in there every week of my life, and you people are out *here?* Who's taking care and feeding the animals?!' Then she proceeded to beat everyone in the line with her purse. Charlie loved to tell that story."

Charlie learned how to work with the unions just as he did every other kind of organization, and later he became instrumental in helping his keepers organize a national association, AAZK, the American Association of Zoo Keepers, so he obviously made peace with the idea and with his union employees.

During those first years with the unions, though, he must have worried about control. How could he be certain the quality was going to stay Schroeder-perfect?

Charlie found a way.

Whatever fears he had, he continued to rule hard and well over the Zoo, and the Zoo continued to flourish.

Tough Taskmaster

"With Charlie, where there was a will, there was a way. And there was always the will," Joan Embery said admiringly, quoting his philosophy: "'It can be done, and it will be done.'"

"He wanted things at the Zoo run his way. He was a loved and feared autocrat," Sheldon Campbell once described him.

"He terrorized the Zoo's employees with his lone prowls," said Neil Morgan, commenting on his infamous memo strolls.

"He was stubborn. He thought he was always right," recalled Bennie Kirkbride, veteran seal trainer and seal show manager. "The thing about it, most of the time he was."

When asked about Dr. Schroeder, one curator during those years began by describing his boss as "an important man, a great man!" Yet the more he talked, the more he'd remember. Finally he said, "But you know, Dr. Schroeder could get mad and just chew on you like you wouldn't believe. He was a real taskmaster."

"There were times when he'd say, 'Oh, you're *going* to do this!' recalled Werner Heuschle, a veterinarian during Charlie's years. "Some of us would tremble a little bit. He was a tough, demanding taskmaster."

The word "taskmaster" seemed to be the description of choice for many who worked for him.

Zoo photographer Ron Gordon Garrison recalled what it was like to make a case with his boss. "I learned early on that Dr. Schroeder was very strong and very opinionated and if your opinion was different, he would listen to you one time. If you could sell your idea, that was fine. If you didn't sell it, then you could come back another time. But for that moment, you backed off. It was either, 'That's right,' and he'd change his direction or 'That's wrong,' and you'd change."

To Charlie, everything had to be thought through carefully, so if things didn't turn out right, it wouldn't be for lack of intelligent effort. Bill Seaton remembers a perfect example:

"One of his favorite questions before any talk or gathering was: 'Bill, have we got a second bulb for the slide projector?'"

It never occurred to Bill, the publicity manager, that a bulb might go out in the projector because it had never happened to him. "But I always had a second bulb for Dr. Schroeder," Bill remembered. "And sure as hell, one night he had no more than said that than we blew a bulb. I grabbed up the second bulb and said, 'I got it, Doctor.' For the rest of my career, whenever I used a projector, I always had a second bulb."

Super-Head of Every Department

On his first day of work as the new manager of group sales, Chuck Bieler had the Schroeder-management style explained to him this way: "I was told that the super-head of every department is Dr. Schroeder," Chuck said. "You found out early that no one thought to be a decision-maker without clearing it through Dr. Schroeder. Every day he had a meeting with a particular department head. He was involved in everything—exhibit design, public relations, marketing, publications, even food and merchandising, since he realized that to grow, the Zoo had to build up sources of revenue. Everybody knew about his little black book, but he also saw every purchase requisition and every piece of correspondence that left the Zoo. He promptly responded to any question or problem."

The revelations would continue for quite a while as Bieler settled into life at the Zoo. The "new guy" found out you couldn't even have lunch without Dr. Schroeder keeping you honest.

"I went out for Christmas lunch with my new colleagues, including Ken Marvin, the head of construction and maintenance. At the restaurant, the phone rang. 'Is Ken Marvin there?' we heard. Ken was handed the phone—it was Dr. Schroeder. He said: 'Ken! What are you doing down there having lunch when you have two buses that aren't running for Christmas week and are going to cost us money?'"

Wow! thought Chuck, deeply impressed.

But Charlie would say this wasn't abnormal. Like any industry, university, or large organization, he explained, you must have departments with principals on top, and the director must have the last word. The director should listen to the input of the total staff, those in which he has greatest confidence, even if he has to draw them out and find out what they have to offer. But he would make the final decision himself.

And everybody knew it. "One time Dr. Schroeder was away," Chuck went on, "and I had a problem that needed an answer, so I went to my boss, Fred Childress. He said, 'You better go ask Shaw.'" Charles Shaw, the Zoo's reptile curator, was also assistant director. But when Chuck asked Shaw, he just said, "Gee, I don't know. Make your own decision or wait till Dr. Schroeder comes back."

Years later, after Charles Shaw's death, when Charlie asked Chuck Bieler to move to Shaw's old desk outside Charlie's office, Bieler was understandably confused. There seemed to be more to the story. As told earlier, he was not to be assistant director but administrative assistant on top of all his other duties.

"But I never had any duties as his 'administrative assistant,' so what did that mean?" Chuck asked himself before finally realizing the answer. "I filled the desk spot, and because someone was occupying the desk, that somehow took the threat away of Charlie having to name an assistant director. The fact is he never did appoint an assistant director. He believed he didn't need one."

Yet Charlie defied the stereotype that might go with a man described in such a way. As Chuck put it well, "Dr. Schroeder was in control and in charge, but he did it in such a way that you never felt he was dictatorial or arbitrary. He commanded that respect."

The evidence of that statement was everywhere. "Even a casual visitor cannot help but notice that Dr. Schroeder runs a tight ship, a Navy term that is particularly apt in view of the number of retired service personnel who can be found on the staff," was the way a 1966 *San Diego* magazine writer described the Zoo in a special 50th Anniversary commemorative issue. "A clean ship is a happy ship," the article said, "and certainly this holds true here, too. Personnel and the animal exhibits themselves obviously have unusually high morale."

Charlie was bigger than life, a force of nature, and those caught up in the whirlwind of such people are often forced to hang on for the ride. If they did, they usually had tales to tell.

"He was a short man, but when he was around, you'd swear he was fifteen feet tall. He was a giant in my estimation," recalled Bob Dale. "He created an image of the Zoo that was so strong, it was as big as San Diego. He was a tough nut who wouldn't settle for anything less than perfection."

"His years as director were the turning point of the institution," stated Jim Dolan, director of animal collections for the Zoo. "He made it the major institution it is in the world today. Everything we have today goes back to that era."

"I have worked for sixty years," said Bill Seaton. "I've had dozens of different assignments from starting a state lottery to being a reporter in Washington, and Charlie Schroeder is on my list of top ten characters. He was so quick and so bright, and he always had these wonderful stories. I have never run into anyone again who was this dedicated, who loved wild environs so much, who was able to cajole hundreds of people, who could make people quake in their boots, who could be so warm and yet so much of a tiger. He did not suffer fools."

Neat Trick

The true surprise of this taskmaster was that Chuck Bieler and Sheldon Campbell were both right: He commanded respect, but he was feared *and* loved. That's a neat trick for any boss. What was the secret?

Charlie was tough, but as is true with the best of leaders, it was a toughness born from desiring excellence for his life's work and demanding the best from those around him.

"He wasn't dictatorial in the sense that most people think of that, but more in the way you expect people to be accountable in order to accomplish a goal," said Joan Embery. "He set the goals, and he drove the team. He was a good guy with his heart in his work, and you knew that he was the reason for the success at the Zoo. Not to get behind Charlie and help him accomplish even greater success would have been crazy."

He might have been the super-head of every department, hovering over everything and everyone, but somehow he also allowed freedom for his staff to spread their wings, another neat trick. He seemed to choose his staff well and then rely on them. "We'd discuss things, then he'd let us work. He really respected his people," remembered Chuck Faust.

"He was organizational-oriented, and he was impatient with people who weren't," said Pat O'Reiley, associate development director who was personnel director during Charlie's years. "And he was also impatient with people who were obviously only here for self-aggrandizement, for what they could get out of it instead of what could be done for the Zoo. I remember the general attitude because many thought that he randomly fired people in piques of disappointment, but in my memory, I never knew him to fire anyone."

Although he obviously had done it at Lederle, he seemed to hate firing people so much that he usually had someone else do it. "Take care of that for me, would you?" he'd say to some unfortunate associate who was standing nearby. In fact, a story is told about a Christmas party at which a keeper had partaken of a bit too much Christmas cheer and pulled a knife in a scuffle with another keeper. The way the story goes, Dr. Schroeder didn't fire him; he just didn't have any more Christmas parties.

There was also something extra about him, though, something intangible that fostered affection for the man because of the infectious enthusiasm he exuded. I was always awed by his power to bring people together with it. Perhaps Marlin Perkins, speaking for his zoo colleagues, captured the quality when he remarked of Charlie: "Don't you feel vibes when you're with him? He's got charisma. We all love him."

Hippies All Wet

Charlie Schroeder, for all his foresightedness, was a man of his conservative times, especially when it came to grooming standards. He once walked into the Zoo during the sixties, and commanded the nearest groundskeeper to turn the front sprinklers on hippies lounging around the Zoo's entrance. His feelings on the subject were so well-known they made a 1969 *San Diego Union* city column. "Hippie-hating Dr. Charles Schroeder informed his employees: 'No beards and long hair allowed here.' He is the director of the San Diego Zoo—easiest place in the world for a beard to go unnoticed," wrote columnist Frank Rhoades.

"He had the strictest grooming rules in the industry. Everyone was always worried about haircuts," said Bill Seaton. "And this was during the sixties, remember. I'd say in front of my secretary, 'I'm going down to see Dr. Schroeder, and she'd say, 'You better get a haircut.' To Charlie, if you had a beard, it was like you had communist tendencies." Many a male employee would hear suggestions about getting a haircut, either by memo or from their immediate supervisor.

But Charlie was also a man who had no problem with some types of social change. When Marjorie Shaw saw an article in the paper about women's business attire now allowing for pants, she routed it to Dr. Schroeder with a note attached asking, "How about us?"

His answer was, "Why not?"

Years before, when top Zoo positions for women were rare and they were having trouble keeping a man in the ZOONOOZ editor's position, in-house newsletter editor Edna Heublein commented that she could do as good a job as any of those men. Charlie heard about it and said, "If you think you can, then go right ahead." After all, he had watched Belle Benchley work all those years, besting most men around her, too. But I suspect that what Charlie liked most was initiative, and Edna Heublein had shown that, plus obvious talent, since she remained ZOONOOZ editor for the rest of her long Zoo career.

A Team and a Family

"We're still working on the sights he set. There is no replacing him," Joan Embery believes. "He was an individual with incredible vision and an incredible ability to bring people together and to understand all aspects of the Zoological Society, which is a very departmentalized, specialized place. Charlie was able to put the team together—and that's what he made it feel like, a team. It was like a big, working family, it really was."

That was the way most people who worked for him felt, especially as time tempers those "hard taskmaster" memories. For an employee to stay twenty-five years was not a rare thing, so he obviously was doing something right to create the camaraderie that kept Zoo employees staying and staying. One prized memo that Chuck Faust, who also goes by "Charlie," kept from those early years echoed that family feeling, even as it grew. There were far too many Charlies working at the Zoo; it was getting terribly confusing. So Charlie Schroeder stopped, made a tally of all his Chucks and sent out a 1959 memo that got it straight:

> To: Entire staff
> From: Dr. Schroeder
> C.E. is quite satisfactory for Mr. Shaw. Hope you are all trying this out. Mr. Faust points out that he has always been known as C.R. so let's stay with C.R. for Charles Faust. As for as I'm concerned, I will be C.S. Let's stay with C.E., C.R., and C.S.

The Three S's

Charlie went out of his way to keep that feeling alive as Bill Seaton remembers it. "Charlie, Chuck Shaw, and I all had birthdays on consecutive days in July," said Bill. "Over the years, we had several parties because of that, and for one of them the staff decided to take us to an outdoor play. They had called ahead and reserved seats, telling the theater people it was a special Zoo party. Of course, back then, the Zoo was number one in the city's eyes, so at intermission someone announced a happy birthday greeting to the three S's from the Zoo: Shaw, Seaton, and Schroeder. Charlie hated surprises, so just as Charlie was starting to gripe about being in the limelight, somebody walked up to us and said, 'I thought that was a little insulting.'

"That got Charlie's attention. 'Why?' he asked.

"'They called you the three asses from the Zoo.'

"Charlie laughed the loudest of us all, and he told that story over and over for years."

The Storyteller

Charlie did love a good story, especially on himself. "He hated meeting with special visitors alone," remembered Bill Seaton. "For instance, when someone like the director of a Polish zoo would drop by, we'd get a call to come down to the restaurant. We'd just sit there and listen to stories. Nothing was ever straight business. I bet I spent half my time working at the Zoo listening to Lederle Labs and Bronx Zoo stories, and you know what? I was always fascinated."

Charlie was a man known for his impatience, and he had a story about that. He received quite a few speeding tickets through the years. He'd tell about the time a highway patrolman pulled him over. He pulled out his I.D. and said, "By the way, I'm the director of the San Diego Zoo," hoping the officer was an animal lover.

"Well," the policeman answered, "if you want to keep that job and not die in an accident, I wouldn't drive so fast." And he wrote the San Diego Zoo director a ticket.

He was a man who never wore anything to work but a dark suit, white shirt, and dark tie, and he had a story about that. Charlie loved to tell how he bought a suit from a Hong Kong tailor on one of his trips abroad—the entrepreneurial kind who would meet the plane and take your measurements right there. A few days later, a beautiful gold suit was delivered to his hotel. On his first day back, he wore it proudly to work, that is, until, as he began his rounds of the Zoo, the seams seemed to give way all at once.

He was, as mentioned earlier, a man who did not like surprises. Oh, how he hated surprises, and he even had a story about it. The Calgary Zoo tried to surprise him once when he was planning a visit to Canada for the Calgary Stampede. A family member was enlisted to ask Charlie, a man who never wore hats, what his hat size was. Charlie came up with a number out of thin air, so at the Calgary Stampede, Charlie was presented with a beautiful, official white Stetson hat that he had to put on—and was four sizes too big.

Whenever possible, he made his feelings about surprises known, as well. Bill Seaton was part of a scheme to deliver Charlie unaware to one of the city's service clubs for an award. "Charlie got up to accept the prize, charmed them with a few of his stories, and then, of course, let me know never to do that again," Bill said.

He also was good at taking as much as he dished out—when the time was right, of course. Chuck Bieler remembers seeing his boss having coffee in a restaurant. "Excuse me, but can I interrupt?" Chuck asked.

"What do you mean, can you interrupt. You already have!" Charlie bellowed.

"Just as I thought I did something wrong, I saw that little twinkle in his eye," said Chuck. "Years later, when I was zoo director, Charlie stuck his head in my office and said, 'Can I interrupt you?'

"'What do you mean can you interrupt me?' I said. 'You already have!' And then, enjoying the look on his face, I said to him, 'Charlie, you don't know how long I've been wanting to do that!'" Charlie laughed as loudly as Chuck did.

The Memorial Wall

As Bennie Kirkbride remarked, Charlie thought he was always right all of the time. But he didn't mind admitting he was wrong...that is, after he was sure he wasn't right.

There's a story told about Charlie's time as interim director for South Carolina's Riverbanks Zoo that its current director Palmer Krantz loves to tell.

"Here I was working with Mister Zoo, so I wanted to do whatever he suggested," explained Palmer. "One day he pointed at a concrete wall around our flamingo exhibit and said we should take it down. He hated it.

"'We can't,' I told him. 'There's a reinforced bar that runs all down the middle.'"

Palmer Krantz thought he'd forget about it. He didn't.

"The next morning, I found him standing in front of the wall with one of my groundskeepers," said Palmer. "I'll never forget the scene I walked in on. Here was five-foot, six-inch, seventy-year-old Charlie Schroeder standing by my six-foot ex-Marine groundskeeper who had a sledgehammer in his hand and the most intimidated look on his face. Dr. Schroeder had just ordered him to whack the wall and he had. Charlie leaned over it, saw the reinforcement bar, looked back at me, and said, 'You were right.' Then he walked away."

The story became so famous around the Riverbanks Zoo that some of the employees nicknaming the patched wall the "Charlie Schroeder Memorial Wall," and is still a favorite story for those who knew Charlie during that time. Krantz, who became the Riverbanks director at age twenty-six after Charlie's recommendation, admits to an affection for the patched place on the old wall.

He remembers Charlie's better lessons, too. "Here was a man who had such an expertise with animals," Krantz remarked. "And yet he taught me a lot about people. He taught me the value of being frank and forthright."

Much More Than a Taskmaster

Perhaps though, there is no more telling commentary than what your peers say about the success of your endeavors and the working atmosphere you've created. By 1966, when the AAZPA conference was held in San Diego in conjunction with the Zoo's 50th anniversary celebration events, Dr. Schroeder received letters from his fellow zoo directors who visited the Zoo. Their comments spoke volumes:

"I am a little discouraged," one wrote. "It doesn't seem to me we will be able to achieve such a high overall level of excellence."

"You have without doubt what must be the finest zoo in the world," wrote another.

"What a wonderful place you have here," scribbled another. "To think that people get paid to work here is almost too much to bear."

Charlie's response?

"What we have done in the Zoo has followed closely the concept of Dr. Wegeforth."

In a letter to Belle Benchley on accepting the director's job, Charlie wrote, "My experience at Lederle has pointed out the importance of avoiding too close business friendships." That sounds cold on paper. Maybe he knew his heart too well. He must have tried very hard to keep that rule in force. Those who worked for him knew the toughness of his "taskmastering," but they also knew the brilliance of that smile and the warmth of his praise. The way he would tell everyone to "Call me Charlie" after his retirement showed an obvious change in some personal policy, once business wasn't part of the equation.

Either way, though, considering the amount of people who, thirty years after his retirement, are still affected deeply by this man is revealing. As often as the word "taskmaster" came up in Zoo employees' recollections of Charlie Schroeder, so did the term "father figure."

Werner Heuschle, former Zoo vet and CRES director, may have described his boss and mentor as a "tough and demanding taskmaster," but the context was a letter to Dr. Schroeder on his retirement. The rest of his written sentiment would echo the feelings of many of those who worked for him:

"I came to recognize and admire your example, the influence of your attitude toward our professional performance and responsibilities, and your fantastic capacity to achieve goals. You were not just The Boss, but a counselor, Dutch uncle, father-confessor, inspiration, and good friend."

"You always knew where you stood with Charlie Schroeder," remembered Bill Toone. "He was like a strict father, but he also had a smile that stretched from ear to ear and nose to chin. And when you made him happy, all you needed was that smile. It was a very rewarding experience, and you knew you wanted to have it again."

Full Gallop

Somehow, Charlie was able to exude all these personas at the same time. Even when working with his Board of Trustees, he had his own way of getting things to work the way he was certain they should.

One longtime Board member commented that working with Charlie was sometimes like "being ants on a log and each of us thinking we were doing the steering because Charlie would have us all convinced the log was our idea. He could get his way, and we wouldn't even know it sometime."

Not that he wasn't beyond the straightforward route if that would work. "He could also be very pointed with the members of the Board, telling them: 'This is needed. This *has* to be done,'" remembered Werner. "I've heard it myself."

Eugene Trepte, former Board of Trustees president who worked closely with Charlie, would agree with that. He saw all of Charlie's wiles and appreciated every one. "Dr. Schroeder was the type of director who was on full gallop all of the time. You were always pulling back on the reins, just trying to stay on," Eugene said. "He had a great sense of humor; he never caused any animosity. You could tell he was a very foxy manipulator. I enjoyed watching him pull off some of these things. Many times I had to cut him off. I'd have to say, 'The Board's feeling, Dr. Schroeder, is that we just can't do it.' But he would never get upset, never show his anger, and by God, he would finally excel."

Other Board members of the time echo the same sentiment, a sort of grudging admiration, even pride, at the firecracker of a director they had on their hands. In a letter written to the Great Plains Zoo director in 1962, former Board president Howard Chernoff, who had many a "meeting of the minds" with Charlie, wrote:

> Charlie is a go-go-go-guy, and the success of our zoo is due to him regardless of what the Board thinks. For example, most people give me credit for our Children's Zoo, but it never would have gotten off the ground if Charlie hadn't conceived it, kicked me and everyone else in the pants to get and keep it going, and then to raise our

> sights to make it first cabin. And he has done just that in many other instances here. Obviously, he has led us to some mistakes, but, as administrators and operators of businesses, you and I know that the guys who do not make mistakes usually don't do very much. He is Messianic on zoos, and it's wonderful that he is.

One Way or the Other

One of the best examples of Charlie's way of getting done what he was deadset on getting done is a story about flags.

During those years, he was very involved with the International Union of Directors of Zoological Gardens.

"Wouldn't it be nice if we had flags in front of the Zoo representing the flags of all the countries of which the IUDZG is involved?" he asked the Board. It would only take $5,000 to put up twenty-two aluminum flagpoles with the flags of each country.

The Board's answer was, in effect, "No, Charlie, we don't think we want to do that. It's a needless expenditure."

Two or three months later, the Board members came to the Zoo for their board meeting and there, in front of the Zoo, were twenty-two flagpoles with flags flying.

The Board members confronted Charlie. They reminded him they had turned the flag idea down.

His answer? "Well, I had this donor, and he wanted to give $5,000 to the Zoo, so I told him about these flags and that's the only thing he would give the money for. What could I do?"

Frozen Flying Sperm

He was certainly notorious for getting people to do things that, suggested by anyone else, would sound crazy.

Bob Smith knows.

Once, Charlie and Dr. Kurt Benirschke, who was the founding director of the Zoo's Center for Reproduction of Endangered Species, (CRES), knew Bob Smith was going to Brussels. The Antwerp Zoo wanted desperately to mate their female pygmy chimp and had been working with Dr. Benirschke to collect sperm from our prolific pygmy chimp Kakowet. Would he mind meeting the assistant director of the Antwerp Zoo at the Brussels airport with a very important "donation" by the chimp?

"I said, 'Of course I wouldn't mind,'" recalled Smith. "So they froze the sperm and put it into a cryogenically sealed container that looked like a milk container. When I got to outbound customs at JFK, the airport customs officer asked to see what was in the container. When I told her I couldn't open it, she asked what was in it. I answered, 'Pygmy chimp sperm.'"

After staring at Smith for a beat, she said, "It had to happen on my watch."

Enthusiastic About Charlie

But if the man was an autocrat, known for his impatient, memo-writing observations of people and their work, he was one with a big heart who was full of awe for the world and capable of enthralled observation of animals and their behavior. That in itself could enthrall others.

"I remember being with him at the Zoo and watching him truly marvel over a new birth he had to show me," said columnist Neil Morgan. "I thought, how dear! I wasn't going to get as excited about it as Charlie, but I was definitely enthusiastic about Charlie."

San Diegan Mary Boehm tells the same type of story. Her parents, Captain William Robert and Grayson Boehm, enthusiastic benefactors of the Zoo, would often invite Charlie and Margaret Schroeder for steak dinner on Wednesday nights at their bayside house. One full-moon night, her parents and the Schroeders were sitting on the porch, watching for grunions in the San Diego surf. Suddenly, all conversation stopped when Charlie saw the Boehme's Persian cat returning from the water looking fierce with a grunion in its mouth. Charlie was captivated. He couldn't stop talking about it.

Mary's parents were struck by how Dr. Schroeder was as interested in a little housecat as a wild cat in the Zoo.

Religion and Science

Characteristically, perhaps because of that enduring, sustaining awe, Charlie had no qualms about tackling, even in print, a big topic rarely tackled by scientists—religious belief. But Charlie, in a Catholic newspaper profile, went out of his way to end the interview with this unabashed quote: "Often religious people feel that science and theology cannot coexist with one another. But I strongly believe that one can look to science and find God. The orderliness in the universe has to be the work of something far greater than ourselves. Therefore science, rather than threatening our belief system, should only serve to reinforce it."

Personal Tragedy

In 1965, personal tragedy hit Dr. Schroeder. His wife, Margaret, died after a long illness with cancer. One of the last trips they made together was a long African adventure. It would be not only a trip to savor because of his wife's condition, but it would also be another catalyst for the concept of a preserve, a wild animal park, which he was already dreaming into reality in his fertile mind.

Their son, Charles Randolph Schroeder, remembers that when they returned from Africa, Charlie would take her along to proposed sites and they would size up each against their African memories. Marge Schroeder was one of the first to see Kenya in the plains of Escondido, thirty miles north of San Diego. After her death, that may have meant more to Charlie than anyone ever knew.

In 1966, a new moated koala enclosure in the Children's Zoo was dedicated to the children of San Diego in memory of Mrs. Margaret Schroeder. The memorial was built with funds donated mostly by her friends at Lederle Laboratories, where she and Charlie had met.

The Director's Presence

Maybe the true test of leadership is how long the person's presence is felt after he or she is gone.

Dr. Schroeder has not been director for close to thirty years and yet every now and then I find myself in discussions with employees who still talk of him in a past-present tense.

Not too long ago, a white koala was born at the Zoo. Dr. Schroeder believed that albino animals were freaks of nature and he would not exhibit them. The current keeper, who could not have been more than twenty-five years old, told me about the koala and then added, "Dr. Schroeder wouldn't have liked this."

She was born after he retired, but she knew her Zoo. Obviously, even now, Dr. Schroeder's lessons and even his taboos are being passed from one keeper generation to the next.

Another example is a Zoo-wide employee survey I instituted. On the question about the director, I received the most incredibly negative responses from the bus drivers, so I called a meeting. *I'm going to do what Dr. Schroeder would do,* I told myself—*I'll just go over there and find out what the hell's going on.*

"So what's this all about?" I asked the group, holding up the surveys.

More than one answered like this:

"I wasn't talking about you, I was talking about Dr. Schroeder." Some of those bus drivers were schoolteachers who had been driving buses since the late forties. The questionnaire had said "Director," and that's who they thought of.

Good or bad, gripes or compliments, the man was larger than life and still continues to exert an influence on us. Even now, at open forums I hold every quarter with employees, someone will raise his or her hand and inevitably ask something that surely would make Charlie flash his big smile:

"Can you walk the Zoo more like Dr. Schroeder did?"

Evening Zoo Walks

He continued his prowls throughout his career, even after his "retirement," but for a while, they were not so lone. After his marriage to his second wife, Maxine, she would often walk with him. The year was 1968, and so much of his handiwork could be seen by then.

Maxine remembers vividly those times with "Charles," as she called him.

"We would go walking in the evening," she remembered. "We lived two blocks from the grounds. So about 5 or 6 p.m., I'd meet him at the back gate of the Zoo. It was lovely. In the evening, the animals change personality. They are quieter, more aware, and very calm. It's a different presentation than the middle of the day when they want to sleep and not be bothered. Something about the time of night made everything different. We'd usually go to the Children's Zoo and hold the animals, but Charles knew the plants as well as the animals."

She saw how he lived with a pad and a pencil, noting that a wire was broken, an exhibit needed repair, or even an animal needed close attention. "Walking that time of evening enabled him to notice animals, to see if they were eating properly, that sort of thing," she noticed. "Charles' dreams for the Zoo were coming to life, and he loved talking about them on those nightly strolls."

Happily Ever After

While jotting down the broken limbs and trash bins in need of care through the years, Charlie's mind was surely wrapping itself around the next big idea to add to his vision. The potential he felt swirling around him must have been exhilarating. A visionary with feet planted firmly on the ground, he must have been inspired by everything he saw as those feet took him through the Zoo's grounds with its special inhabitants. During his first years, those feet must have taken him often by the new home of four friends from Lederle Labs.

If anyone doubted his respect and loyalty to the animals who gave so much to research of human diseases, one of his first acts as the San Diego Zoo director showed the heart he had for the creatures who helped us conquer some of the modern plagues of our time.

In 1954, at the height of the poliomyelitis epidemic, four special chimpanzees came to live at the Zoo, no doubt through Dr. Schroeder's grateful gesture of an invitation.

"Four VIPs hopped off a plane at Burbank yesterday en route to a life of ease in San Diego," stated the front-page news story. "The VIPs were Buck, Peanuts, Banjo, and Pete—chimpanzees who had made medical history. They are still immune to polio, although they got their shots two years ago. Chimpanzees are lifesavers to medical researchers, for they are the only members of the animal kingdom, outside of you and me, who can contract polio," the reporter went on. "They had been the pets of Lederle Laboratory at Pearl River, New Jersey, where they served as guinea pigs in the developing of a polio vaccine. The four chimps, after a brief stopover, went onto the San Diego Zoo in Balboa Park for a life of ease as a reward for the medical services."

Perhaps Buck, Peanuts, Banjo, and Pete's presence kept his mind on his first zoological love. For all his natural marketing savvy and business acumen, Charlie Schroeder continued to be a scientific man devoted to research. It was where his true passion lay, where his thinking always returned.

While Charlie was ramrodding sky trams and moat-construction, pushing his staff to new heights of excellence, charming city councils, Board members and the public, he was no doubt brainstorming the transformation of a place tucked in a corner of the Zoo's grounds, the place where he'd first fell in love with the Zoo and animals like Buck, Peanuts, Banjo, and Pete—the Zoo's hospital and research facilities.

CHAPTER 14

The Other Side of the Zoo

From his first day as director, the animal care and research part of Charlie's "Zoo plan" was certainly never far from his mind. A year after arriving in 1955, he had already written an article for the *Bulletin of the San Diego County Medical Society* called "The Other Side of the Zoo."

"Tucked away in a remote corner of the zoo, hidden by the Old Globe Theatre and well-isolated from the crowds of sightseers, stands the two-story Zoo hospital, a monument to one-time Zoo director Dr. Harry Wegeforth and well-known philanthropist Ellen Browning Scripps," Charlie lyrically began the article.

"In the matter of zoo hospitals, San Diego ranks with the best. The extensive facilities are used both for research and for teaching. Research may be carried out by outsiders, but projects must be approved by the research committee. Are you interested in research?" he added. "The physical facilities of our building are at your disposal." In 1927, with Ellen Browning Scripps' funding, Dr.

Harry built what was originally called the Zoological Hospital and Biological Research Institute, hoping it would grow into its name. When young Charlie Schroeder, D.V.M. came along to help in 1932, the new hospital was Charlie's first domain, and he was, in a very literal sense, its first mover and shaker.

Now he had a chance to finish the moving and shaking. Dr. Wegeforth had envisioned the need for research in a zoo setting; Dr. Schroeder breathed life into it as Zoo vet, and then later as Zoo director. In a real way, he never strayed far from that remote corner of the Zoo.

"Recreation, education, research"—those are words on the Zoo's original charter. One of the stated by-laws is "to actively engage in biological research." Milton Wegeforth, Dr. Harry's son and Board president during Charlie's first years, summed up his father's thoughts about Charlie's efforts: "Dr. Wegeforth would have sided with Dr. Schroeder with respect to research because research is going to be the answer to propagating endangered species."

Of course, "siding" may not have been the word Charlie would have chosen—"blessing" from his mentor probably would have been closer. Charlie never forgot the letter, mentioned earlier, in which Dr. Harry made clear that his whole ambition was to have "the zoo revolve around the research. I do not care a 'tinker's dam' to merely run a circus. . . ."

As the new director of one of the biggest zoos in the country, Charlie Schroeder could have easily become so busy that he'd let some old research habits rest.

But as the man of business forged ahead, the man of science was still writing papers. The in-house newsletter *Zoo Bell* has this 1956 entry revealing for the topics mentioned:

"Our director Dr. Schroeder will be attending the Conference of International Union of Directors of Zoological Gardens to present two papers: one on antibiotics in the zoo, and the other an application of industrial management practices in the zoo."

He was a man comfortable in both worlds and never became too busy to add to the universal discussion of both worlds. As soon as his money management and marketing ideas began to pay off with a black-inked bottom line, he began making plans to upgrade the veterinary hospital and department.

A Good Idea Waiting

From the day of his return, Charlie realized that the Zoo's vet department, and zoo medicine in general, was not up to his speed. The new discipline he'd helped pioneer was still a small one. After all his years in the upper echelon of animal research, his first veterinarian goal at the Zoo was to have what he would call "a legitimate operation." He began to bring in people for his hospital, some he would later call "frauds," as many were unable to live up to his quality expectations for animal care, so Charlie began another old practice from the thirties. He was not shy about getting outside help from the city's medical community to save animals. As he worked to bring up the quality of zoo medicine by hiring and training more Zoo veterinarians, he began calling in "people doctors," as had been his practice as the Zoo's vet years before, charming them into volunteer consultations.

Obviously he passed that wisdom on to the veterinarians, who ultimately would make up an excellent hospital and research staff. By 1971 Dr. Charles Sedgwick would be quoted in a local paper about the continuing practice of using medical doctors whenever possible: "When we have a problem beyond our experience, it's always possible to call in men from the medical community for help," Dr. Sedgwick explained. "For instance, University of California at San Diego physicians are helping us design an intensive care unit."

Then, to Charlie's certain approval, he made a definite point to praise in print the volunteer pediatricians, orthopedic surgeons, and internists who continue to come to the aid of the Zoo "always without compensation."

Whales and Innovation and Roosters

While Charlie wasn't shy about calling on outside expert medical help for animal health problems, he himself had a reputation as someone to call for such problems. Charlie had acquired so much experience as a pathologist while in the commercial field that he had become a very good technician.

"He was sharp. I'd go to him with a problem first thing," agreed Bennie Kirkbride, the Zoo's longtime sea lion show trainer. "He could tell you, if anybody could, what you needed to know about sea lions."

Even Sea World during its first years called Charlie for advice, which prompted Charlie to call one of his favorite "people doctors" to help.

"One of their first whales was sick and wheezing," said Dr. Homer Peabody. "Charlie called me in for consultation with the whale because theoretically, I am supposed to know why people breathe, so obviously I am supposed to know why a whale is having trouble doing so. After a huge amount of pondering with Charlie over what to do with the wheezing whale, we decided to take an X-ray. He had some problem in his wind pipe or breathing tubes that was making the noise, but how were we going to get a X-ray of it? Charlie had an idea, and it was a good one. He said, 'Why couldn't we use one of these big huge industrial pictures for detecting flaws in cement?'"

Charlie just happened to know of a place in El Cajon that had such a machine, and immediately went to investigate. Next thing Dr. Peabody knew they were back with a very big machine that could take four different picture segments of the enormous mammal. "With this huge X-ray picture, we mapped the whole whale lung and saw something in the left lower lobe," Dr. Peabody recalled. "The whale had aspirated what looked like a Coca-Cola bottle cap. I remember Charlie saying, 'They just don't inhale something like that. I bet what happened was some kids were flipping bottle caps, aiming for the blow hole.' We knew then we'd lose the whale since there was no way to operate to take the thing out. And that's what happened. We tried to save it, but it developed obstructive pneumonia, which did not improve following massive doses of penicillin. All these years, though, I have been impressed with Charlie's innovation. He was very clever."

Charlie also fielded questions not in need of innovation, and he could never resist sharing a good story with the newspaper, such as this entry that found its way into Neil Morgan's *Tribune* column: "A local bank president has been dabbling in poultry farming. He has some chickens which have been laying eggs. But he's been unsuccessful in getting any of the eggs to hatch, so he consulted Dr. Charles Schroeder, San Diego Zoo director.

"'Is there a rooster with your hens?' Dr. Schroeder asked.

"'What does a rooster have to do with it?' the banker replied."

Fantastic Laboratory

As he prodded the Society's Zoological Hospital and Biological Research Institute into a new era, Dr. Schroeder's own convictions about not only the importance but the obligation to research the exotic animals in his care never wavered.

He put it well for a newspaper interview during that time in typical Charlie fashion: "A zoo is, in fact, a fantastic laboratory containing more than a thousand different kinds of animals, each anatomically and physiologically different," Charlie said. "Variations are genetically brought about through natural selection in their particular environment. This is hardly a new story, but is beautifully demonstrated in a zoo. Examples are all around us. There is the anatomical adjustment of the giant anteater, for instance, which permits it to penetrate holes and crevices in search of food with its modified tongue supplied with an especially tenacious saliva to collect insects and deposit them where they will do him the most good.

"Then there is the peculiar ability of the koalas to find all nutrient requirements in the leaf of the eucalyptus," he went on. "Even its drinking water comes from the leaf. Every animal in the zoo is unique in its adjustment to its environment. Though the research is for animals, there may be benefits to mankind."

In a scholarly article in 1968 on the research potential of zoos worldwide, he stated his feelings with unusual emotional scholarly fervor: "Research efforts to date have been puny, and for the most part, studies which might give greatest reward have not been pursued."

Genetics, especially cytogenetics and reproduction, ecology, general physiology, health-preventive medicine, use of antibiotics and related drugs, should be studied in depth, he stated emphatically. Very few zoos at the time were research zoos, or as they would be called later, "scientific zoos." Charlie must have felt a deep passion for San Diego to one day be among them and even take strides ahead of the rest. He would help make that happen in a very short amount of time.

Attitude of Education

One of the problems in developing a top zoological hospital and research center was the same one as it was in the thirties. The specialty was not as small as it was when Charlie began his veterinarian career, but it was still a new discipline, which meant enthusiasm would be more plentiful than experience in those he hired. Education, then, would be the way to the staff he believed worthy of the Zoo.

"He was proud of the caliber of both staff and visiting scientists who carried on their studies at the Society's Zoological Hospital and Biological Research Institute," wrote *ZOONOOZ* editor Marjorie Shaw. "He was devoted to Scholia, a scholarly group to which he belonged and for which he recruited many others. He championed writers that interpreted science for the lay public, and he always required his staff to be aware of research elsewhere. Curators at the Zoo during the fifties and sixties were constantly encouraged to keep up with the scientific journals in fields other than their own," she wrote.

"Even if they didn't have time to read scientific papers, they were expected to peruse tables of contents to become familiar with titles and subjects of work. Stacks of journals, magazines and correspondence were routed to staff members daily. There was scarcely desk room to accommodate the material. But Dr. Schroeder had gone through it all, and so could they."

When Charlie was awarded the Zoological Society's Conservation Medal in 1976, he was lauded for having created an "attitude of education."

Institute for Comparative Biology

Soon Dr. Schroeder had a Zoo veterinarian staff he could rely on, including a trained pathologist, and he had a research committee committed to sharing his dream.

In 1962, the National Science Foundation awarded $100,000 for the purchase and installation of laboratory furniture and equipment to upgrade the existing structure, followed by thousands more in modernization. The laboratory floor space more than doubled the existing 3,000 square feet already housing the hospital and quarantine area.

Before Charlie was through, $650,000 was sunk into transforming the complex into the highest-caliber modern research hospital and laboratory for its day.

In 1963, Charlie changed the name of the Zoo hospital's Biological Research Institute arm to the Institute for Comparative Biology, and he pumped new life into his first dream. He opened the doors and invited scientists to come work with him at the Zoo for the betterment of animals and humans alike. There was much to learn.

"Some day the world may refer to the happenings here in a relatively small building on the San Diego Zoo grounds with the same respect accorded the events on Noah's Ark," wrote a *Los Angeles Times* science columnist on the Institute's opening.

By the end of his director years, a reporter would describe the foreseeable impact the changes Dr. Schroeder would set in motion: "Partially hidden by trees, by the Fine Arts Gallery and the Old Globe Theatre, the San Diego Zoo Hospital scarcely ever is seen by the public, but its work affects us all." The establishment and nurturing of the Institute for Comparative Biology was the important first step toward the creation of the Zoo's prestigious Center for the Reproduction of Endangered Species that was formed by Dr. Kurt Benirschke, a man whom Charlie was ecstatic to have as part of the Zoo's "scientific family."

Two kinds of research were established at the Institute of Comparative Biology—research done by the staff for the Zoo, and outside research performed by graduate students and professors from universities with specific approved research studies. His goal was to offer everything: comparative animal behavior, biochemistry, pathology, physicology, parasilogy, bacteriology, and virology. Viruses began to receive immediate attention from the Institute, and an animal virus data center was established. Soon the collected and discovered viruses were being examined a part of an international survey.

Charlie enjoyed boasting of the continuing evolution of the virus studies of the Institute. "We have the only comparative virus laboratory in the world right in San Diego. It's a new field. There's a population explosion. There are probably 200 viruses lying around out there. As we get smarter, we're finding many, many more viral diseases we are only now discovering."

The money was tight, but ideas helped spread the word. Life-appointment fellowships were awarded to renowned biologists. The research committee sponsored quarterly seminars to which the fellows were invited to hear papers. The Ellen Browning Scripps

Foundation's grants-in-aid established in 1939 continued to be given. A summer student program was established, which funded fellowships for students pursuing human and vet medicine, some offering a year as a Zoo intern.

"One is trying to track down a leukemia-lined virus," began a new profile on a group of summer interns in 1963. "Another wants to find a basic diet for animals. The third is trying to ride an ion through a medical loop. Who? Salk? DeBakey? Crick? No, the names are Karp, Kuehn, and Holloway. These three young students, two males and one female, were from Stanford, Minnesota, and Dallas, and they are spending their vacations, not on the beach, but working hard at the San Diego Zoo Hospital."

The funds were low, but the incentive and the facilities were attractive to all sorts of research from many sources. The federal government funded an investigation of hibernation by deep-space exploration experts to determine whether humans could be placed in a state of suspended animation for space voyages. San Diego dentists collected and studied teeth as the animals shed them. A study of five bears who died from the same kind of cancer was undertaken.

The results were noticeable within a handful of years. In 1972 Dr. C. E. Cornelius, dean of the University of Florida Veterinary College, already had these glowing words to say about Dr. Schroeder's impact: "Needless to say, the many doors you have opened for biomedical research endeavors are beyond counting. The success of an able administrator is always weighed and related to how much freedom of activity is possible under his direction. The many productive research experiences we have all had working at the Zoo attest to your ability to make the Zoo something very unique in America," Dr. Cornelius wrote. "My personal activities at the Zoo over a fifteen-year period are dear to me and relate directly to your visionary role in acknowledging the possible role of the public zoo in contributing to the comparative biology and medicine."

Zoo's Autopsy Room

"A young and eager reporter was overheard in a United Nations pressroom asking an Arabian health delegate: 'Sir, what is the mortality rate in Arabia?'

"My good man, it is 100 percent. Allah wills that we shall all die.' And so it is in the zoo," Charlie wrote in a 1955 article for the *Bulletin of the San Diego County Medical Society.* Since 1929 every death, from Charlie's viewpoint, was a lesson in waiting for those with eyes to see and the equipment with which to see it. From the very beginning, he was determined to have up-to-date equipment and to continue to train his doctors to discover and document what they found.

Because of Dr. Schroeder's longtime interest and belief in animal pathology, the Zoo hired a full-time pathologist in 1964, then designed an autopsy room beyond compare for the time. Its autopsy table merited a news story all by itself in 1971. You can almost see Dr. Schroeder picking up the phone to his latest editor best friend to brag about it. Pathologist Lynn Griner designed it for animals too big for regulation tables.

"A table made of iron pipes sits on rubber tire wheels," wrote Zoo reporter Betty Peach. "Overhead is a track with a crane, capable of lifting a two-ton load. On the table lay a white bearded gnu, victim of a jealous male that had torn down a fence to attack his opponent."

Soon Dr. Griner would create, in a pre-computer era, something called a "Termatrex"—a punch-card system making the data on 10,000 autopsies he alone had performed in order to make the data quickly retrievable and transferable to computer storage when the time came. That, in a real sense, was the essence of "comparative biology," gaining knowledge by comparing information gathered from all the health experiences occurring daily in the world's biggest Zoo.

There were two distinct phases of Dr. Schroeder's new "health department" staff's work—the prevention of disease and treatment of sick animals, and a systematic compiling of the scientific knowledge to be gained from every animal death. And behind it all hummed the research labs.

As Zoo research hospital manager Bernard Sheridan "proudly and accurately" described the place to a *California Veterinarian* magazine journalist in 1967: "There is nothing like it anywhere."

New Ark Beginnings

By 1975, that quote would be changed to "You ain't seen nothing yet" as the Center for Reproduction of Endangered Species (CRES) came into existence as the full-fledged, trailblazing research department for the San Diego Zoological Society. The Zoo's research committee was headed by Dr. Kurt Benirschke, a physician who conducted pioneering work in human twinning. A professor of pathology and reproductive medicine at University of California, Dr. Benirschke was a medical man with a passion for endangered species. He, along with other members of the committee, decided to take Dr. Schroeder's vision and challenge the Board of Trustees not only to "get serious about doing research here," but to also make a difference in the mushrooming problem of saving endangered species in new ways.

In visionary words that might have been written by Dr. Schroeder himself, the research committee's proposal prepared by Dr. Benirschke stated that "we believe that the possibility now exists for our Zoo to become the leader in this field, as it has in exhibition and is evolving to be in conservation. We believe that the idea is viable and represents one of the most necessary next steps in the development of the Zoo."

The step was taken.

Dr. Schroeder's efforts had turned the Zoo's attention toward this aspect of Zoo life, slowly and surely. Today, CRES is in existence due to what those at the Center have called Dr. Schroeder's "far-sighted planning and unflagging energy," and that is certainly true. While Dr. Schroeder would say much of his work was to put legs on Dr. Harry's dream, Charles Bieler and Dr. Benirschke were furthering that dream in the same visionary way.

Now with a full research staff, studies are conducted in endocrinology, behavior, virology, and genetics. Dr. Benirschke's "frozen zoo" begun early in his career became the nucleus of the Zoo's now-famous Frozen Zoo, which is arguably the biggest and oldest now in existence. It would be a perfect compliment to the sudden, vast opportunities that the Wild Animal Park's wide-open natural spaces offered in captive breeding.

It would take a decade of work and a decade of world change to culminate in CRES's creation, but all through that time, the shot of adrenaline Dr. Schroeder gave the Zoo's fallow research arm continued to expand and grow. San Diego joined the small list of "scientific zoos" in the world, and now has taken a place among the best. Dr. Schroeder would be proud. The "remarkable future"

Dr. Wegeforth predicted when he opened the original Zoo hospital is here in ways no one could have foreseen.

Dr. Benirschke, who remains a dominant figure in that remarkable future, knew Charlie before moving to San Diego. He tells a little-known story that underscores Charlie's influence in the zoo world and his dramatic ability to make things happen for revolutionary new ideas—his favorite kind.

In the sixties, Dr. Benirschke became good friends with Roland Lindemann, who owned the Catskill Game Farm in upstate New York where he worked on projects with his hoofstock. Dr. Benirschke noticed that the staff was losing quite a few animals. They were not waking up after immobilizations using curare, for which there is no antidote. "In the mid-sixties," recalled Dr. Benirschke, "that was a major problem in animal treatment, to immobilize them but make them get up. I persuaded the staff at Catskill to try a new drug used in Africa that was being manufactured in England called M-99, a morphine analog, with an antidote that much of the rest of the world was using, but was not legalized here," Dr. Benirschke said.

Through an international colleague, Dr. Benirschke was able to acquire some of the drug. They used it to immobilize a zebra, a guanaco, and a llama, all of which promptly woke up with the antidote. Dr. Benirschke filmed the procedures to show other animal care professionals in hopes of saving more animals by using the drug.

And that was where Charlie came in. Dr. Benirschke was in San Diego working toward a research grant with the Scripps Foundation Committee and had written Charlie for permission to show the film to his hospital staff. Charlie, of course, was intrigued and promptly agreed.

"I challenged the vets to see the drug as the anesthetic choice of the future for sedating animals," recalled Dr. Benirschke of the day he showed the film to the Zoo's hospital staff. "I remember they didn't take me seriously, even laughing at it, but Charlie did. He supported my efforts to import and legalize the drug. Now it's used everywhere. That success is due in no small part to Charlie's backing it and using his former relationships through the chemical industry. M-99 has been modified, and new drugs have come along, but it was the initial drug that offered this life-saving option." The rest of the story, of course, is that by 1970 Dr. Benirschke had moved to his new professorial position with UCSD's University Pathology Department and Reproductive Medicine. Charlie had promptly asked him to be on the Zoo's research committee, and soon the story would come wonderfully full circle when Dr. Benirschke founded CRES.

Dean L. K. Bustad of Washington State University's Veterinary College, Charlie's alma mater, summed up Dr. Schroeder's contribution nicely in a commemorative issue of *Zoo Animal Medicine* celebrating Dr. Schroeder's eightieth birthday. "The Zoo was more than a business to Charlie. He developed an outstanding educational program, which has been emulated and copied by many other zoos throughout the world. He was totally responsible for modifying the zoo's animal hospital." Bustad wrote. "His philosophy is that zoos should be a learning experience involving all living things. Charlie feels that zoos are very important cultural and scientific institutions that should serve as repositories of information about wild animals and as breeding centers for endangered species. He believes the research effort should go beyond the use of exotic animals for medical research and should be concerned with perpetuation of rare animals in their natural environment and in captivity."

As New York Zoological Society's Ross Nigrelli put it even better in a letter to Charlie on his retirement: "You may be 'Mister Zoo' in San Diego, but to me and to your numerous friends in the scientific world, you are an outstanding scientist in the field of comparative medicine. You have shown the zoo world how a 'menagerie' can contribute to basic knowledge in science and medicine, in addition to being a place for education, recreation, relaxation and for the sheer enjoyment of seeing rare and exotic animals."

To See Is to Protect

Dr. Schroeder, Scientist, never fell into the "ivory tower" syndrome. He knew the importance and the power of being an educated zoo lover. He took to heart the words of conservationist Sir Peter Scott, who once told him, "Charlie, if you want people to protect something, they've got to see what it is they're protecting."

Charlie agreed. The more the public knew about "the creatures

CHAPTER 15

A Conservation Turning Point

In 1926 Dr. Harry Wegeforth went on an expedition to the Guadalupe Islands, about 250 miles from San Diego, for one purpose: to find Guadalupe fur seals, which were thought to be extinct. Dr. Harry heard that some had been seen, but he returned home without having any luck.

On the way back, though, he spread the word at each landing that he'd pay handsomely for anyone who'd bring the Zoo a pair. On April 25, 1928, a fisherman arrived in San Diego with two live and healthy Guadalupe fur seals. It was one of the exciting moments of Dr. Harry Wegeforth's life as a zoo man. During a time when wildlife was plentiful and ripe for the taking, that must have been a moment that awoke a feel for conservation before it carried any of the modern apocalyptic weight.

"Our interest in preservation of species leads naturally to our endeavors to discover new species and to throw light on those already known," Dr. Harry wrote in *It Began With a Roar.* "Our most singular reward for this has been our discovery of the fur seal on Guadalupe Island. There is nothing which has attracted to this Society the attention of scientists and statesmen more than the appearance of these beautiful creatures."

Within a handful of months, Charlie Schroeder arrived at the Zoo to be the Zoo's new veterinarian. Was Charlie impressed with the fur seal discovery? No doubt he was. Charlie became well-known for his amazingly early understanding of the shrinking wild and its ramifications for the future. When did Charlie begin looking at everything not just through a research lens but also through a conservation and captive-breeding filter? Perhaps his turning point was much the same as Dr. Wegeforth's. Perhaps it was his first close encounter with the endangered wild itself.

September 1961 was the month that Charlie took his first trip to Africa. He and his wife, Margaret, had their own moments of never-forgotten communion with the wild. He wrote a five-page, single-spaced Christmas letter to friends about that first trip to Africa, and his famous capacity for awe served him well in the way he expressed some of those moments.

"The dream of our lifetime was realized when we went on a six-day safari in Kenya, East Africa, to see animals we were so familiar within and without their natural habitat," he wrote. Along the banks of the Athi River, he was awed by stepping in the huge foot-prints of an elephant and "never knowing when we might meet face to face."

At a lake in Nairobi National Park they saw gerenuk. "These shy, delicate, long-necked creatures were among our favorites," he said, "and we had thought they were quite rare, but we saw at least sixty." He wrote about seeing a herd of three dozen elephants, all sizes, from tiny babies up, moving out of a line of trees into an open meadow. And he wrote of beautiful creatures with orange heads and turquoise bodies—geckos—who kept them company at night at the safari lodge, squeaking to one another. "I'm sure if it had not been for our mosquito nets they'd have shared our beds."

They met some tribesmen and saw their way of life. "All this reserve is Masai reservation, and we saw many of this colorful, handsome tribe herding their scrawny cattle," he wrote. "During the recent severe drought and locust plague, it is estimated they lost 50 percent of their herds."

And then, on their way to the Park's lodge, they and their driver became completely lost in a fierce dust storm. "Casting about trying to find a traveled track, we came up on three lionesses who were as lost and bewildered as we were. A few minutes later, we saw a magnificent male. Just as we were getting anxious, the wind dropped and through curtains of dust there was Kilimanjaro, which had been hidden by clouds all day. There are moments in everyone's life, times that are profoundly moving, and this was one for me."

By 1963 he would visit Africa again, this time to attend the meetings of the International Union of Conservation of Nature held in Kenya, and he would return to write a *ZOONOOZ* article, complete with his own photos of the wild, entitled "Conservation Is Our Business." He became quite aware after the African conference of the coming crisis of habitat that would hit Africa's wild treasures. The world was no longer looking like a world of plenty, even at that early date.

Charlie ended the article with this pronouncement:

> Conservation broadly embodies all natural resources. At a zoo, however, where we are concerned primarily with animals and plants, our immediate interest is in the preservation of the fast-disappearing exotic species of birds, reptiles and mammals around the world. The Zoological Society of San Diego can and will disseminate information on wildlife preservation and continue its efforts to secure and exhibit pairs of animals we are trying to save, and we will attempt to induce these vanishing species to reproduce in captivity. Animals for this purpose will be secured only with the full knowledge of those charged with conservation in the countries of origin on the recommendation of the International Union for the Conservation of Nature and Natural Resources, and special cooperating committees and organizations.

It sounded like a rallying cry, and in many ways, it was. Conservation became the topic he began to speak on, the gospel he began to preach. Regulations and laws, such as the Endangered Species Act, didn't go into effect until the mid-seventies.

By that time, the Wild Animal Park was already established and on its way, but Charlie had already been beating the conservation drum for over fifteen years. For instance, in a speech he gave in 1965 to the Pacific Coast Oto-Opthalmological Society, his topic was the importance of zoos in conservation efforts. He began by quoting London Zoological Society's Sir Solly Zuckerman: "No one wishes to contemplate a world whose surface had been entirely transformed into centers of population and industry, into areas of mechanized agriculture, and into endless stretches of concrete roads."

What role can zoos play in conservation efforts? Charlie asked.

"Zoos can stop traffic in rare animals, the bootlegging and falsification of legal documents," Charlie said. "This can be done by encouraging legislation in the countries of origin, and prohibiting the capture and shipping of rare animals unless accompanied by a bonafide permit. Those zoos which do acquire and exhibit rarities can make every effort to have pairs and attempt to bring about reproduction," he went on, all but pounding his bully pulpit. "In this way, zoos can become the source of rarities for other zoos, rather than depleting numbers in their native habitats."

This was to be the foundation of his Wild Animal Park vision and passion, and it would remain his topic of choice for most of the many speeches he would give for the rest of his life, as was the case in 1983 in his commencement talk given at his University of Washington's Veterinary School:

"The most productive way to perpetuate a species is to protect it on its home ground—its natural habitat and not in a zoo," he said. "Eliminate man-made incursion, especially loss of territory, is the direction to take. Reproducing rare species in captivity is a commendable approach, not to repopulate a threatened area, but to learn about reproductive behavior and populate other zoos without recourse to capture in the wild."

Zoo-50

Meanwhile in April 1966, the Zoo had a party like San Diego had never seen.

The Zoo was fifty years old, its Jubilee Year, and it was time to celebrate. Sheldon Campbell, longtime Zoo devotee and future Board president, was appointed chairman of the festivities, and the planning went on for months before the event. It was a time set aside, as *ZOONOOZ* coverage put it, "for looking at a colorful history, savoring the present, and anticipating the future of the Zoo."

The future, Charlie believed, would be conservation. With all the planned fun, so was some very serious exposure for the role the San Diego Zoo wanted to play in fostering that future.

It was up to Sheldon Campbell to define that exposure, but even Campbell knew the true head of his Zoo-50 "department" would certainly be Dr. Schroeder.

What could the Zoo offer to reflect the scientific trailblazer mentality among the cake and the parades and the frivolity?

Why not a conference on conservation? And along with the conference, why not the establishment of an annual Conservation Medal to be given annually?

On top of that, why not have a speaker who would not only be a scientific feather in the Society's cap, but also do a whale of a job at packing in the public to hear the message?

Campbell suggested a young woman named Jane Goodall, who had just made an enormous public splash through a *National Geographic* special about her study of chimpanzees. Something about her work and persona had struck a chord with those who had seen the *Geographic* special on her work.

They invited her, and she accepted. The newspapers were alerted:

"Final plans have been completed for the San Diego Zoo's birthday party, which begins at 10 a.m. tomorrow when Baroness Jane Van Lawick Goodall will cut the first piece of a birthday cake that is three feet wide and 100 feet long," read the article announcing the big day. It would be billed as the world's largest children's birthday party. After all, the Zoo was held in trust by the city for the children of San Diego. The cake was designed to serve the 30,000 children expected. Actor Marshall Thompson, star of the hit TV series *Daktari* about a jungle veterinarian, would make an appearance. Plus the film *Born Free,* based on Joy Adamson's best-seller of lions in Kenya, would have its West Coast premiere in San Diego to raise funds for new exhibits at the Zoo.

It was going to be a *big* week.

The International Conservation Conference, however, was set for the fall of that year. Since the April week of festivities would come first and be noticed more by the press, Charlie wasn't initially in favor of Jane Goodall speaking at the celebration, claimed Sheldon Campbell. With all her exposure, she struck Dr. Schroeder as a bit too popular, Campbell said, so Charlie questioned whether she had credentials needed to impress the scientific world.

If that's true of Charlie, then the scientist was again warring with the savvy marketer. Her vast popular exposure obviously had the same effect on him as *Zoorama*. He loved the idea of the exposure, but he worried about the level of scientific credibility for the honored speaker.

The balancing act was on yet again.

A compromise was struck when the Zoo co-sponsored an invitational seminar on primate behavior that same April week with the Western Behavior Sciences Institute and San Diego State College's Division of Life Sciences. The seminar paired Goodall, the expert on chimpanzees, and UC Davis's anthropologist Phyllis Jay, a solid scientist with years of Indian field studies of langurs.

So Charlie got his headline: "ZOO WEEK TO FEATURE 2 WOMEN SCIENTISTS."

Charmed by Jane

Charlie may have had his reservations about Jane Goodall, but Campbell claimed he was soon "charmed by her just as everyone else."

Charlie's anecdote about that night was this one:

On the evening of her big speech, he and Campbell escorted Goodall to the front of the city's Convention Hall, which was packed to capacity with people eating their Zoo Jubilee "safari supper" and waiting to hear her speak. Her husband was the projectionist, showing movies of her work with the chimps, so he was off setting up his equipment.

"How are things going?" Charlie asked her.

"I'm really nervous, just terribly upset," she answered.

"Well, don't you eat that very poor chicken dinner you've got there," Charlie answered. "We will take you out for a nice dinner after this is over. So just rest, take it easy, and you'll do a great job." And that's exactly what happened.

"The crowd went into ecstasy," said Charlie. "Sheldon Campbell and I took her out for a big steak dinner, and she ate it with gusto."

How did Charlie describe Goodall after the fact? "She was a young behaviorist on her way up, not world-renowned yet, but she enthralled the crowd."

Months later, still celebrating the Zoo's 50th year, the conservation conference would be held with all sorts of world-famous conservationists, including Sir Peter Scott.

"We had a great variety of research endeavors," said Dr. Schroeder, the solid scientist. "We held it in the same place, but didn't have one-fifth the turnout we had with Jane Goodall," noted Charlie, the publicity whiz.

What We Did to the Dodo

"The whole concept of the nature and function of the Zoo is changing, and it has to change. If those who are most closely concerned with wild animals—the zoos, universities, research institutes, game departments—do not all fight for conservation, then the wild animals which are their work and their interest, their justification for their existence, will in many cases cease to exist."

So said *International Zoo Yearbook* editor Caroline Jarvis, one of many voices brought together to talk about this new, suddenly all-too-timely subject of conservation. The place was San Diego, the occasion was the International Conservation Conference, and the date was October 4-6 of that Golden Jubilee year, 1966. The theme was "The Role of Zoos in the International Conservation of Wild Animals."

The conclave may not have attracted as big a crowd as Jane Goodall, but those who participated were the ones who would shape the future of the world's zoos, and by extension—whether anyone liked it or not—possibly the very existence of the wild and its creatures. The loss of the wild and its endangering of species upon species seemed to become epidemic overnight to all but the handful of conservationists who had kept "crying in the wilderness" until they were heard. The conference was designed as a place for all those cries to be heard and heard well, where perhaps meaningful change could be fostered. The stated aim was to "stress the role that zoos can play in conserving the rare species of wildlife by providing protection within the zoos, by promoting governmental restraint on hunting, and by encouraging governments to set up game preserves."

Prominent conservationists from Australia, Thailand, England, Germany, Sweden, South America and South Africa attended. Charlie made sure all his fellow International Union of Directors of Zoological Gardens (IUDZG) members were invited. Chairman and lead speaker of the conference was Commander Peter Markham Scott of the World Wildlife Fund, and those attending read like a Who's Who of the conservation movement.

Other principle players of the zoological world included Major I. R. Grimwood from Peru, Dr. Boonsoon Lekagul from Thailand, Perez M. Olindo from Kenya National Park, Harold Coolidge of Switzerland, and Ian Player, chief conservator of Zululand, South Africa, who played a decisive role in saving the Southern white rhino and who would be very much a part of the Wild Animal Park saga.

In a nice touch, a member of an extinct species also made an appearance. The last passenger pigeon, named Martha, who died in 1914 at the Cincinnati Zoo, made the trip across the country from the Smithsonian Institution where she is displayed.

With Martha's presence, endangered species were singled out for discussion in hopes that some effect could be made on their future. The Galapagos turtle, polar bear, Arabian oryx, African cheetah, orangutan, Mongolian wild horse, Kashmir stag, grizzly bear, Java rhinoceros, giant panda, American bald eagle, California condor, Hawaiian duck, ivory-billed woodpecker, and whooping crane were all showcased as being on the verge of lost forever.

Martha's presence alone would move one reluctant Board member to change his vote on the existence of a Wild Animal Park—an ongoing source of dissension and stubborn persistence since Charlie first brought the idea of a second zoo campus to the Board's attention in 1959.

But that's a whole other story to be told next, one that reads like a suspense tale and ends with a legacy.

Peter Scott's keynote speech, entitled "A Voice Crying in the Wilderness?" is moving and timely even now and worthy of a look more than thirty years hence:

> What man did to the dodo, and has since been doing to the blue whale and about 1,000 other kinds of animals, may or may not be morally wrong. But the conservation of nature is most important because of what nature does for man.

I believe something goes wrong with man when he cuts himself off from the natural world. I think he knows it, and this is why he keeps gardens and window-boxes and house plants, and dogs and cats and budgerigars. Man does not live by bread alone. He should take as great pains to look after the natural treasures which inspire him as he does to preserve his man-made treasure in art galleries and museums. It has been argued that if the human population of the world continues to increase its present rate, there will soon be no room for either wildlife or wild places, so why waste time, effort and money trying to conserve them now?

But I believe that sooner or later man will learn to limit his own overpopulation. Then he will become much more widely concerned with optimum rather than maximum, quality rather than quantity, and will rediscover the need within himself for contact with wilderness and wild nature.

No one can tell when this will happen. I am concerned that when it does, breeding stocks of wild animals and plants should still exist, preserved perhaps mainly in nature reserves and national parks, even in zoos and botanical gardens, from which to repopulate the natural environment that man will then wish to recreate and rehabilitate. All this calls for action of three kinds: more research in ecology, the setting aside of more land as effectively inviolate strongholds, and above all, education.

Even if I am wrong and man reaches the stage when there will be standing room only on this earth—even then the conservation effort will have been worthwhile. Many will have enjoyed the pictures even if the gallery is burnt down in the end.

An Announcement Denied

You can almost see Charlie nodding his approval from his seat by Sir Scott's podium, and you can almost see his disappointment, too, because Charlie wanted desperately to make an announcement at this conclave, an announcement about such a place with just such a bold commitment right here, right now.

He had even mentioned the possibility of the announcement in a recent newspaper interview, no doubt "to help the cause," as Charlie was known to do, but Charlie would be denied his moment.

A *San Diego* magazine article during the celebration summed up Dr. Charlie Schroeder's place in the Jubilee celebration: "The modern-day San Diego Zoo actually began thirteen years ago when Dr. Schroeder took office," the article proclaimed. The modern Zoo that Charlie built would continue to transform beyond all imagination, but Charlie had one more idea up his sleeve, one he knew would be his legacy, one he knew this 1966 group would have stood and applauded. While Charlie Schroeder was transforming the San Diego Zoo, behind the scenes, as far back as 1959, he was forming and fighting for a dream.

Charlie envisioned a zoo of the future with ten times the space and ten times the capacity for saving animal species, a zoo that would populate the world's zoos without disturbing the wild, and a zoo that would raise human awareness of the crisis.

Yet it would take all of Charlie Schroeder's political savvy, all his network of connections, all his formidable charm and clout—and even a few Charlie-styled tricks—to finally, finally, make it happen.

Section 3

CHARLIE SCHROEDER

Stubborn Visionary

"Animals need space, to love and to run,
to explore new areas, to climb and do all the things
that come naturally."

—Charlie Schroeder, Wild Animal Park creator

CHAPTER 16

—The Battle for the Park—

Charlie's Dream

1959

The scene was always the same. High on a hill overlooking San Pasqual Valley, thirty miles from the San Diego Zoo, Charlie Schroeder would stand and share what he alone saw with his companions of the day.

It happened every few weeks. Charlie would bring a friend, a dignitary, a colleague, a city official, a group of Zoo patrons, a celebrity, a trustee, and usually a bucket of fried chicken. He would stand on the hill, spread his arms as if to take in the whole surrounding valley that was nothing more than scrub brush, cactus, and rattlesnakes, and he would share his dream.

Over there, he'd point, will be the East African valley. Over here will be Asian plains. And right there, a watering hole designed exactly like one in the South African wild. The animals will roam free—rhinos, antelope, giraffes. People will see them all from a train winding through the preserve, having an experience they could never have anywhere this side of a transcontinental adventure in the wide-open wild.

Can you see it? It will work! It will be fantastic!

"There was nothing here," recalled Jim Dolan, the Zoological Society's director of collections. "It was a farm, a cow ranch with dried cow poop on the ground. Charlie would stand up with his hands raised and tell them what he saw. He had it all plotted out in his mind."

"He had such a knack for getting people to believe in what he believed in," said Suzanne Strassburger, former Park public relations director.

"Charlie was the world's best salesman," stated John Fairfield, one of the first Wild Animal Park animal keepers. "He had the ability to paint a verbal picture of something that didn't exist anywhere but in his mind, yet he would fill your mind so full with it that you'd either want to jump in and help him build it or pay money and have somebody else do it so you could see it quicker."

Some were dazzled by the acreage, the valley, the scheme; others thought Charlie Schroeder was crazy. But almost everyone in Charlie's path during those years got an earful of his vision.

To look down from that ridge today is to have your breath taken away by scores of rhinos and giraffes and cheetahs and antelope and cranes and hundreds of other species grazing below in a certified botanical garden of 3,200 rare and endangered animals.

What did Charlie see?

He saw the zoo of the future.

The animals would roam free over hundreds of acres. Creatures made rare through their vanishing wild homes would have a space to reproduce and survive.

People—rather than animals—would be enclosed. Visitors would view them from a small train winding through the preserve, and Charlie believed they would willingly drive to this remote spot and pay to view it all.

Critics, including members of his Board of Trustees, said it couldn't be done. The Park would be a failure and financial drain on the Zoo. Over and over, Charlie kept bringing it up for a vote; over and over, they voted it down. One Trustee even threatened that if Charlie brought the matter up again, he'd vote to fire Charlie.

What did Charlie do? He brought it up again. In fact, the threat deepened Charlie's resolve to continue pursuing his grandest dream at a time when most men retire. Over the years, as he continued transforming the Zoo in Balboa Park, he kept making his fried-chicken treks with influential San Diegans to that hill above San Pasqual Valley.

In 1965, Bob Smith and Neil Morgan were taken up on that ridge. Twenty years later, Smith recaptured his feelings in a letter to Charlie:

> Looking around the 360-degree sweep of it all, you traced the raw, undeveloped landscape where you envisioned Asian plains, Asian swamps, East Africa and South Africa. I remember how enthralled both Neil and I were at the time. Had we not known that you were the man who could make it happen, who would know how to draw on the support of every person and organization you would need in order to get the job done, we might have thought it simply a dream.

A *San Diego Evening Tribune* editorial would state that the opening of the Park marked not only a new recreation area but "the start of one of the boldest wild animal preservation projects in the U.S."

Charlie has been called "a prophet with Noah-like vision," a fitting accolade because the Wild Animal Park can also be called a Noah's Ark for the 21st century. Other "game farm" concepts for profit would be attempted, such as the Lion County Safari Park chain, which cropped up around the country in the seventies. They are gone, but the nonprofit San Diego Wild Animal Park gains more praise and attention each year for its vision.

As was his gift, Charlie collected the ideas swirling around him as he transformed them into something brand new. The Wild Animal Park is Charlie's gift at its most quintessential. The result is a zoo oasis in high-desert country so original and so trendsetting that public agencies, private enterprises, and nonprofit organizations around the world line up to learn from it.

In fact, the San Diego Wild Animal Park is so unique that even now we strain to explain its essence. Which Park do we describe? Charlie's dream was an evolving one, and the Park is so many things today:

- a breeding ground to populate the world's zoos without depopulating the wild
- a new kind of zoo where animals live in large spaces and feel more freedom than any zoo in existence
- a conservation ground where the people in this place, at this time, do all they can to save species from extinction.

Yet it almost didn't happen.

In fact, without Charlie's unflagging, impossibly stubborn, do-or-die will, the Park would have never been built—without any doubt. Who else would have weathered years and years of effort, a hail of negative votes, and a threat to his job for even the most wonderful of dreams?

Charlie Schroeder, a man famous for the speed with which he got things done, worked on this dream for the better part of a decade. He never gave up; he never took a no vote for an answer. In truly Schroeder style, just as he had done with all his other big concepts, he made plans, got his staff involved, and then began to push his vision, complete with practical evidence for its success to the Board.

Jim Dolan remembers Charlie in full stride: "I don't think Dr. Schroeder ever thought anything he was involved in would be anything but a success; that was the mentality of the man."

This race would take nine years, five votes, and some fancy Charlie Schroeder maneuvering to keep the dream alive.

But that's getting ahead of the story.

How and why Charlie began to dream this specific dream is the place to begin.

Revolutionary Idea

By 1959, Charlie had already began envisioning the Park with people like Chuck Shaw and his other curators. "We wanted to create a big area that was compatible to the animals in the hope that they would reproduce," he once said. That was the simple idea.

The bigger idea was this: It was time for us to quit scavenging the wild for our zoos. "His idea was to try stop depleting wildlife populations, to establish self-sustaining captive populations between zoos," recalled Jim Dolan, who would have full responsibility for populating the Park. "In that respect and at that time, the idea was really farsighted. Zoos were still importing large numbers from African and Asian countries, where regulations were not in place. His idea was that we should only draw from wild populations on occasion for new genetic materials. For the most part, Charlie believed we had an obligation to maintain what we had in the best possible condition, and to reproduce them for the long haul. That was a revolutionary thought process."

From a zoo director's point of view, at least one as farsighted as Charlie Schroeder, the very existence of animals and zoos alike was at stake. Both could die away in just the span of an animal's life. If we cannot stop the wild from vanishing with the encroachment of humanity, then where would zoos find their animals? That was the practical question. Charlie, however, was asking two more: If we *can* stop the wild from vanishing, shouldn't we? And if we can stop a species from vanishing, how can we not come to its aid?

Those were the visionary and responsible questions.

So Charlie stepped into his bully pulpit, stoked up his formidable charm and drive, and began to preach about the future he saw and the dynamic impact he believed the Zoo could make on it.

To Love and Run Naturally

Charlie also seemed to have another reason for his passion about an open-spaced animal park. In an interview years later, we can hear the underlying emotion he had for the subject in his words:

> Our gorilla exhibit at the Wild Animal Park is one of the largest anywhere. But it's not enough. Animals need space, to love and to run, to explore new areas, to climb and do all the things that come naturally. There isn't enough room for that. When you speak of the zoo in Balboa Park, there is nowhere for them to go. There are ninety-two acres locked in. It's bigger than zoos used to be. Let's face it, there were times when you had a tiger in an area sixteen feet by eight feet deep. Anybody knows that's not adequate for a tiger, and there lots of zoos that have beautiful tiger exhibits, and they're pretty big, but they are inadequate.

> No, the Park is not even enough, but we've tried. The idea of putting animals in their natural setting is not new, but the presentation is. The free-ranging animals move in herds, not in pairs, as in most zoos.

Where did Charlie get the idea for the Wild Animal Park?

The old way of running a zoo was to have as many different species as possible and to showcase rarities that other zoos did not have. As Theodore Reed, director of Washington's National Zoo, explained the zoo world reality during those years, "Every city zoo emphasizes variety. We keep more than 700 species here. That means small enclosures. We can do pretty well with cats and bears and small mammals, but hoofed animals are like cattle. To breed them you need herds. What zoo can keep thirty giraffes or elands?"

Only a few zoos tried to answer that question and create a new zoo philosophy. These were Charlie's colleagues and in a few cases, his mentors.

Gerald Durrell, who opened a zoo on the Channel Island of Jersey, had begun preaching that zoos must become "stationary arks."

Roland Lindemann's Catskill Game Farm in upstate New York had been raising rare hoofed animals for years. His 1,064-acre preserve bred many endangered species such as white-tailed gnus, Przewalski's horses, and Pere David's deer. Washington's National Zoo and the New York Zoological Society both established breeding reserves. In Europe, Walter van Den Bergh, director of the Antwerp Zoo, oversaw its wild animal park at Planckendael; the London Zoo had established Whipsnade, forty miles outside the city.

Most of these outlying preserves were not open to the public. Charlie spoke often of visiting a man named Jimmy Chipperfield who had begun an unusual game preserve. "Jimmy had made a deal with the Duke of Bedford to use a portion of the Duke's 3,000 acres of land," said Charlie. "He divided it and put lions in one section and cheetahs in another, baboons in another. Then he had one big area where you would drive in with your car and you'd pay just one price for everybody. You'd bring your lunch and you'd sit on the ground with the giraffes and the rhinos and the antelope."

Charlie's first plan for the Zoo's game preserve was even less complicated. Initially, he and his staff envisioned a turnout from the nearby road, where cars would pull up, face in and look out at the animals. They foresaw creating containments for the animals, drilling a water well, and building a snack bar and restrooms, all for under a million dollars.

As the Board continued to vote down the idea, however, Charlie's concept, instead of withering, began to evolve into the complex place of wonder it is today.

The beginning, though, was one simple idea—a game farm.

CHAPTER 17

The Battle Begins
1960-1963

When Charlie and his curators realized they needed more room, they began to hatch their idea of fencing in the surplus stock, letting them do what comes naturally, and with the results of such captive breeding, populate the Zoo, as well as others, in order to leave the wild alone.

In 1960 Charlie decided to approach the Board with the idea of the simple game farm—but not before he had rounded up the same group that had worked so well planning the Zoo's moated enclosure transformation as well as the Children's Zoo. As Chuck Faust tells it, Charlie would call a senior staff meeting, hand each of the men a clipboard and a pencil, then give the command that they were there to consider how to start a wild animal preserve somewhere.

"Anybody who had an idea had to write it down," Chuck remembered. "For weeks and weeks, it was a regular thing, and we were getting quite a bit accomplished."

He even took them along to look at different pieces of land. "Dr. Schroeder would always show up in his best suit and tie," Faust recalled. "He'd stop the car, and we'd get out to see where the boundaries of that particular piece of land were. Some of us would start out going around a hill, but he'd start up the mountain—coat and tie and all—in the middle of summer, and we all had to follow him. He did that so many times to us, it was hilarious."

Minutes-Keeper

And as the Schroeder legend goes, Charlie had another thing or two going besides good planning. Charlie set the agenda for each Board meeting and kept the minutes of each Board meeting. By all accounts, he took fine advantage of both of these circumstances. Supposedly, if Charlie believed something needed to be left out or adjusted a bit, he was not above doing so, and that went double for each meeting's agenda.

In a 1983 *San Diego Union* interview, Charlie said he not only loved creating the Wild Animal Park, he enjoyed the politicking that went with the job: "Recalling how he got his way with skeptical zoo trustees, he grins a satisfied grin, leans forward with a conspiratorial air and asks rhetorically, 'Who do you think wrote the agenda for each meeting?' He throws back his head and shakes with laughter. 'Then I'd go right into my office after the board meeting and dictate the minutes—the whole thing!'"

That would explain a lot. If ever an account was documented by a positive thinker, it was this one, and it's fascinating reading for anyone who has ever heard Charlie's legendary way for making things happen.

The Setting

The Board of Trustees met in a room that housed Belle Benchley's big mahogany desk, an ancient beauty designed to be used by two people. The chairs were in a circle around the desk, and the Trustees sat informally in them. The president of the Board would sit behind that desk, and Charlie would sit beside him in a large chair with papers in his lap and a second chair nearby. Charlie would bring up a subject, and the Board members would discuss it. Then he'd make notes on the top piece of paper and lay it on the second chair.

According to the minutes on December 26, 1960, Charlie brought the new idea of a "back country zoo" to the Board.

"It was announced that Dr. Schroeder and his staff members met with city official Henry Clay. The Zoo's proposal to use county land for a game farm was greeted with enthusiasm," wrote Charlie the Minutes-Keeper. "The Society may have as little or as much county land as they wish to fence." There was even an island in the middle of the San Vincente Reservoir area where the Zoo wouldn't need fencing, Chuck Shaw reported.

The game reserve idea would be explored by "staff," said the historic record.

According to Charlie's minutes, the record also states, rather emphatically, that there was no opposition to the concept.

Charlie was on his way.

This Is It!

Then something happened that must have excited Charlie tremendously.

He must have been talking about his idea with everyone because one day right after the new year, the phone rang. On the other end was fellow Rotary Club member Clayburn LaForce, who also happened to be the city employee in charge of leasing city land. "Charlie," he said, "I've got just the place you're looking for."

North of San Diego, the city had just built Sutherland Dam. The farmers down through the valley sued the city for water rights. They believed the dam would make the water table fall and harm their livelihood. They won the case, and the courts ordered the city to buy the land below the dam. The city did, turning it into an agricultural reserve called the San Pasqual Agricultural Preserve. LaForce was the man who leased the agricultural land back to the farmers.

Part of the preserve, however, was a section of land surrounding the historic San Pasqual Battle Monument that the agricultural preserve could not totally use. Did Charlie want to come take a look at this patch of land on the "wrong" side of the Highway 78? Charlie explained the moment in this way:

"I picked up the assistant city manager, our superintendent of construction, and a curator, drove up to the valley, looked it over, and immediately said, 'This is it!' Clayburn was being paid fees for the leases, so he didn't intend on giving us all the land since 200 acres of the land was plantable and he expected to lease that for crops," Charlie went on. "But I got so excited that when we went back to see Clayburn and his maps, I drew straight lines from the property to the highway and said, 'Too bad Clayburn, we're taking it all!'"

In January 1961, Charlie, curators Charles Shaw and George Pournelle, and a city water department official took Carl Hubbs, the Zoo's exhibit committee chairman who also happened to be a Board member, to see the land.

The next month's exhibit committee minutes read:

> Thirty-one miles north of San Diego near the San Pasqual Battle Ground, an area ideally suited to a display of many of the threatened species of ruminants has ideal cover interspersed with open areas to allow good visibility. . . .The area has the advantage of being adaptable to species with different ecological requirements. The area consists of three basins ringed with hills. This project was enthusiastically approved and will be pursued further.

At the Board meeting two weeks later, Chuck Shaw made a formal illustrated presentation called "Wildlife Area in San Pasqual Valley" about an uphill track of over 1,800 acres available through the water department of the City of San Diego on a long-term lease basis for a nominal fee. A majority of the Board voted for "management" (which meant Charlie and Chuck Shaw) to prepare a formal document. By the next Board meeting, the same majority voted for the go-ahead to explore a lease on the property, and a committee was formed to negotiate the lease consisting of board members Anderson Borthwick, Laurence Klauber, and Howard Chernoff.

Charlie must have thought the dream was on its way.

Africa in His Eyes

In September of that same year, 1961, Charlie and his wife, Margaret, took their first trip to Africa, and Charlie returned with

East Africa in his eyes. Now he had a sight to go with his dream. He must have come home and gone straight to the hill overlooking the San Pasqual land and marveled at the similarities.

After that, he began to explain the Park very specifically. It was to be "a little Kenya, a living diorama, where herds of animals can roam naturally in terrain resembling the African highlands," as he told *California Veterinarian* magazine after his return.

He also came back with a vastly deeper passion for conservation, and that began to be part of his evolving Wild Animal Park vision that he shared with one and all.

First Strike

But then came the first big problem.

Board president Howard Chernoff worried that leasing such land might be in conflict with the Zoo's charter with the city, so the Zoo's lawyer, Sherman Platt, was consulted. The answer was yes, it might be. The charter's wording stated that the tax of two cents on every $100 of assessed valuation of city property—set up by Dr. Harry Wegeforth at the Zoo's inception—was "to be used exclusively for the maintenance in Balboa Park of zoological exhibits." From the lawyer's view, a game preserve way out in the far-reaches of the county might only be leased using normal Zoo income after a charter amendment.

Also the Board—knowing the tenacity of their director and the dangers of expansion too fast and too soon now that the Zoo was finally doing well financially—had agreed that the decision to build any such park should be unanimous.

When Charlie returned from his African safari, the Board greeted him with their decision and with big questions.

The October 1961 Board meeting minutes read:

> The Back Country Zoo—in the course of discussion it was suggested that policy-wise the Trustees should determine whether it is:
>
> 1. a good thing
> 2. feasible
> 3. require change of Charter

Purpose to be clearly defined:

1. conservation
2. breeding farm
3. establishment of a second Zoo

Move was made to table the program until outside funding is available.

Strike one against Charlie.

The issue would not be brought up again for almost a year and a half. The Board was just understanding this "go-go-guy" didn't quit. Only four years earlier, Charlie had won a hard-fought battle for the Children's Zoo, and he was still building moated enclosures throughout the Zoo as the wire came tumbling down. This time, however, they had to slow him down. Good idea aside, if the charter had to be changed, this might be too much too fast.

As Board member Eugene Trepte put it so vividly, the Board members were always pulling in the reins, and for the moment, they were pulling hard, but Charlie had seen Kenya, and he had seen it in the San Pasqual Valley land.

Was the idea dead in the water?

For now, it was—to everyone but Dr. Charlie Schroeder.

High Flight

Months passed, and then a year.

In February 1963, while Charlie's friends in the city government waited to see if the Zoo would go through with the lease, Charlie decided it was time for another run at the Board. This time, however, the exhibit committee would do the presentation, not Charlie. Better yet, the chairman of the committee would do it—alone.

Of course, that happened to be a Board member named Carl Hubbs, world-renowned UCSD marine biology professor who would later be the namesake for Hubbs Sea World Research Foundation. The minutes read: "Chairman Hubbs made a case for the back country Zoo to reinterest the Trustees in the project."

Would the Board allow further investigation, Hubbs asked? Would it allow a study to be done to answer the questions that halted the progress a year ago? The Board voted yes.

So the exhibit committee went back to that hill overlooking the 1,800 acres of San Pasqual Valley.

Then one of the members went a step further.

G. C. Ewing got into his plane, flew over the site and, when he came back to earth, sat down and wrote this letter to the Board:

> After this afternoon's visit by the exhibit committee to the proposed site of the back country zoo, I took another look at the site from the air. Now I am more convinced than ever that the place constitutes a marvelous opportunity to develop a facility that will, I think, be unique in the United States. The location is very strategic.
>
> As the county population grows, it will be very accessible to a large group of users. Such a project would, of course, have great practical utility to the Zoo and would be of very far reaching scientific interest to ecologists, botanists and conservationists.
>
> In addition, because of its spectacular setting and because of its unusual nature, it cannot fail to become one of the outstanding attractions to residents and visitors in San Diego. At the moment, the climate seems favorable to undertake this venture at little cost. Therefore I enthusiastically recommend that immediate steps be taken to insure that we shall not let this opportunity slip away.
>
> Sincerely,
> G. C. Ewing
> Member, exhibit committee

If Charlie Schroeder was not in the passenger seat of that plane, which would be a very good guess, he was definitely there in spirit. He had another convert. In fact, the whole exhibit committee had caught his dream.

A Master Plan

But what about the city charter problem?

Professor Hubbs met with the city's attorney to work out some sort of a solution. His companions on the trip? Charles Schroeder and Charles Shaw. They came back with the news that the charter was "flexible."

That's exactly what they told the Board.

While questions remained about the charter, the Board voted to allow the lease to be secured, and as soon as it was, to allow a master plan to be prepared.

The date now was April 1963. Charlie didn't want to wait for a lease to be secured before continuing his pitch for the Park because no one knew how long that would take. After all, considering the city charter brick wall still in the way, where would the money come from to pay for a lease?

Nowhere fast, Charlie surely knew.

Perhaps the master plan should come first, he must have thought. *Perhaps that would shake some money loose and push the lease along without all this city charter fretting.*

Besides, while a master plan might take quite a chunk of time for some people, it wouldn't for Charlie Schroeder. In the same good habit he learned from Dr. Harry Wegeforth, Charlie and his staff already had most of the information prepared and waiting.

When the Board members arrived for the next meeting, they were told the master plan was being mailed to each member, listing the financials and even the number of antelope species earmarked for immediate residence on the game farm.

Professor Hubbs went on the record as speaking not only as chairman of the exhibit committee but also as a conservation representative of the Zoological Society. He was convinced that "the charter provision under which the Zoo operates would not prohibit the use of areas for such purposes as holding temporary surplus animals, for breeding animals for exhibiting in Balboa Park, and for cooperation with other zoological societies in helping to prevent the extermination of threatened species." Besides, he deftly pointed out, the Board had already agreed some years back to fence another section of land elsewhere for such purposes.

"I strongly urge approval," Hubbs said. "The Society should immediately enter into negotiations for the lease while every effort will be made in the interval to find construction funds from private and public government sources."

Out There in the Boonies

Then some very good news came from downtown: the San Pasqual land had been annexed to the City of San Diego. Charlie, Professor Hubbs, and the whole exhibit committee hoped that would help. No longer was this just a county project; now it could be a city one. One staff member at the time recalled that there was "very strong opposition to doing that crazy thing out there in the boonies." The land now being part of the city could help fight the "boonies" perception, as well as assuage other possible political and financial questions raised about such a project outside the city limits.

Why did the city annex the property? The story goes that the water rights was a cover story, but the real purpose was for the Zoo. City Manager Tom Fletcher stated later that he felt the city needed such a project to promote the North County area.

Charlie had another convert, and a very important one. As one friend of Charlie's explained his style: "He didn't go to the San Diego City Council. He went to the one guy. The power structure was very small, and a phone call could get something done in those days."

Charlie's powers of persuasion were formidable, so I can believe almost any such story. He could con us all into doing the most amazing things, including picking avocados from his backyard orchard. Without mentioning avocados, he'd say, "Doug, why don't you come on over to the house this afternoon if you've got time?"

"Sure, Dr. Schroeder. Be happy to," I'd respond.

"Doug, do you like avocados?" he'd say when I arrived, handing me a bag. Then for the next hour, we'd pick and talk, and when I left, he'd hand me two avocados and say, "There you go." So I'm never surprised when people who knew Charlie before and after his retirement tell me similar stories about being caught in his web of persuasion.

But whatever was the truth, with or without the help of Charlie's charm, the city annexed the property out there in San Pasqual Valley and waited for the Zoo to make the next move.

Charlie and the exhibit committee announced the good news to the Board and waited to see if this development would push the Wild Animal Park proposal over the top.

It didn't. Neither Professor Hubbs' eloquent plea nor the city's land annexation broke the Trustee deadlock.

What was really going on?

In a news article, Board member Robert Sullivan explained it this way: "There was no opposition to the idea, but there was a lot of feeling that it would take the Zoo's money and spread it too much."

It's not too hard to guess the logic of the Board's resistors. The Zoological Society was just now beginning to emerge from under the financial straits it had always found itself in. The money was rolling in but flooding out to pay for all the Schroeder-inspired improvements throughout the Zoo. Good idea or not, how could they waste money on a second zoo?

Strike number two.

CHAPTER 18

"If You Bring It Up Again ...!"

1964-1965

That fall, Charlie took his second trip to Africa, this time to attend the meetings of the International Union of Conservation of Nature held in Kenya. Charlie came home with another burst of energy for his game preserve dream. His vision was continuing to evolve, and so was Charlie's resolve.

To stir the waters again, Charlie finagled a meeting with the city attorney, city manager Tom Fletcher, the director of public utilities, and this time, a new Board member believer, Minton Fetter.

Afterward, at the February 1964 Board meeting, he and Fetter reported that the newly annexed land was still being set aside for the development of a zoo conservation and breeding area by the city. Then the minutes added this: "The city manager pointed out that another use will be found for the property if the Zoological Society Trustees decide against the use of these lands."

Perhaps Tom Fletcher, the city manager, thought a little pressure might help. It didn't.

Yet again, the Board decided not to move forward.

This had to be strike three. You can almost feel the disappointment from reading Charlie's dictated minutes.

Stuck into the official account is a letter from Minton Fetter to city manager Fletcher urging him to give them a few more weeks.

"I am personally very enthusiastic about the possibilities, and I believe a majority of the members of the Board are favorable. However, in the interest of harmony, I do not believe a decision can be reached by March 17th," he wrote.

"If you could do us the favor of informing the city council that you have tentatively set this area aside for use of the Zoological Society conservation program, I believe we can eventually convince the entire Board."

No one knew weeks would turn to months, and months would turn into years.

Those Who Resisted

Who resisted Charlie's dream?

Board members such as Howard Chernoff, Gordon Gray, Laurence Klauber, the renowned reptile expert and namesake of the Zoo's reptile house, and Frederick Kunzel, U.S. District Attorney, were the ones who seemed most worried about Charlie's rash new idea, but Klauber and Kunzel were the most vocal. The two strongly opposed for economic reasons, according to Sheldon Campbell, who joined the Board later. "They believed a wild animal park would get too involved with itself, would not represent the quality of the Zoo, and would be a financial drain on the Zoo," he explained.

Successful, conservative pillars of the community, they loved the Zoo, and some had been with the Zoological Society through many of the same financially strapped years that Charlie had experienced back in the thirties with Dr. Wegeforth and Belle Benchley.

These men, who had lived through the Great Depression, did not warm to the risk of crazy dreams and impractical schemes. Imagine their view of this wild man of a director who was already transforming the entire Zoo.

How do we talk common fiscal sense to this man? they must have wondered. *What's all this crazy talk about a second Zoo?*

The Hell You Will

Charlie recounted an outburst by Fred Kunzel during one of the more heated Board discussions about the proposal: "Charlie, you can't do that. You'll wipe out the reserves!" Then Charlie remembered the moment that Kunzel, soon to be a U.S. District Judge, went ever further. Concerning the fees and the use of Zoo income, Kunzel suggested they make it a test case in court.

"'The hell you will,' piped up a couple of the others, thankfully," Charlie recalled. "But Fred was right about the city funds," he admitted. "The city charter stated that every two-cent tax on every $100 tax evaluation was written into the city charter for the maintenance of a zoo in Balboa Park, so we couldn't use those funds for a wild animal park purpose. We had to find another way."

To Skin a Cat

Hindsight is perfect, of course, and it's hard to imagine why anyone wouldn't vote for such a grand scheme as today's Wild Animal Park. Consider, though, what the Board must have been dealing with in those days. CRES deputy director and longtime friend Andy Phillips grasped the probable situation behind those closed doors well:

"Can you imagine being on the Board in the 1960s and hearing each new addition to Charlie's Park idea? 'We're going to develop 1,800 acres, and it's going to cost us millions of dollars, and we're going to put in a monorail, and we're going to put in this huge water line, and we're going to be nonprofit!' Charlie's dreams were never small ones."

Board member Eugene Trepte reminisced about those meetings and Charlie's force of personality:

"Charlie would never show his anger, which was so fine a part of his character. He would think, *Well, if I can't make it with this approach, I'm going to try another approach.* He always had many ways of skinning the cat and eventually, one of the approaches got through," Trepte explained. "That was the type of individual he was and why we made so much progress. A lot of leaders would've thrown up their hands and said, 'Oh hell, they said no! Okay, if they don't want it, I won't do it!' And nothing would get done. But that isn't the way of a great institution. Instead, Charlie just said, 'I believe in this. They should have confidence in me, but they won't go along with me, so I'll come up with another approach.'"

General Victor Krulak, who joined the Board in 1969, remembers how important poise was to Charlie. "He couldn't be seen as one who lost his temper," Krulak said. "You'd say your piece, he'd say his, and you would disagree. Then you'd say goodbye. Yet you knew he was going to take another run at you, and you knew you were going to take another run at him."

That was certainly true. "Irrepressible" was a word that could have been coined for Charlie Schroeder. He knew who his adversaries were, and he certainly understood their love for the Zoo. What he could not understand was why he couldn't change their minds.

He already had three strikes against his idea, yet he refused to be called "out." Only a man like Charles Schroeder could keep swinging.

Helluva Politician

Frank Curran, San Diego's mayor at the time, remembered a few of the complications surrounding the game preserve concept. "We had a little bit of problem because the valley was designated as an agricultural preserve," he told a reporter. "But finally we figured that if you could raise cows on one side of the street, you could raise elephants on the other side." He also remembered Charlie Schroeder as "bullheaded."

"Nobody got in his way," Curran added. "He was a helluva politician, too."

He must have been because it would be three long years before the lease—the one that Minton Fetter asked the city manager to hold a "few weeks"—would be officially contracted between the Zoo and the city. Until that moment, however, he had to keep his dream alive and his dream valley on hold.

But how, Charlie must have wondered, *do we do that?*

The answer?

A study.

A red-ribbon, objective, official, university-connected, impossible-to-refute feasibility study.

Yes, Charlie must have decided, that just might do the trick. He went back to the Board.

Stanford Feasibility Study

The idea for a feasibility study was promptly passed by the Board's willing majority. The minutes read: "The director was told to contact the Stanford Research Institute to produce a feasibility study for the 'so-called game preserve' and answer specific questions that the Trustees devised."

By the way, the Institute was not to give opinions. Policy decisions would be up to the Board.

You can almost hear the skeptical Board members wondering, *How do we keep Charlie away from these guys?*

But if that was the case, Charlie let it ride. He seemed to have faith that his idea would hold up to the outside scrutiny from experts. As Jim Dolan put it, he never considered that anything he believed in would be anything but a success.

Is Otto the Answer?

Meanwhile Charlie began to work on finding money. If financing was to be the major hurdle, it was time to try to solicit funds from private or public agencies, or maybe even outright donations.

Remember Elmer C. Otto and his $1.5 million bequest during the mid-sixties? No doubt, the timing was perfect in Charlie's mind — the Elmer C. Otto donation had materialized from nowhere and was big enough to fuel several big ideas. Charlie must have perked up, and he must have also expressed his hope to a few of his newspaper friends because the very first mention of Charlie's dream found its way into Neil Morgan's newspaper column in 1964:

"The $1.5 million Otto bequest to the San Diego Zoo will likely be spent on capital improvements for our city's No. 1 tourist attraction," wrote Morgan, "including an education center for school students at the Zoo, and a Zoolandia: a conservation and breeding farm being considered for two square miles of city land at San Pasqual. Decisions are expected in thirty days."

Can you imagine the reaction at the next Board meeting? Which by the way, was in thirty days.

Charlie Schroeder single-handedly laying out the proposed Wild Animal Park's monorail line, 1970

Early visitors watching the Wild Animal Park under construction

Early African-inspired signs marking the site of the Park-to-be

BEFORE: *An aerial shot of the Wild Animal Park site, 1969*

AFTER: *An aerial shot of the Wild Animal Park only a few years later*

The view from the Wgasa Bush line monorail on an inaugural trip, including Charlie, Arthur Godfrey and an interested giraffe

The view of the monorail from the residents' point of view

BEFORE: San Diego Wild Animal Park's Nairobi Village under construction

AFTER: Nairobi Village, a veritable botanical garden, years later

After 23 days at sea, South African's Southern white rhinos are released to their new home on the California plains, 1970

South African conservationist Ian Player and Charlie view the thriving Southern white rhino herd on Opening Day, 1972

The Wild Animal Park's early elephant show with Joan Embery

Today's Benbough Ampitheatre, home of the Park's free-flying bird show

A typical view across today's Eastern African enclosure

A Kenya impala herd running across the Park's wide-open spaces

The Congo River Fishing Village, inside the Park's Nairobi Village

An unusual view of the unique place these Wild Animal Park giraffes call home

Charlie and Maxine Schroeder at his retirement party attended by 700 of his closest friends

Director emeritus Dr. Schroeder being honored for the Wild Animal Park's success by Zoo director Chuck Bieler and Wild Animal Park general manager and future Zoo director Doug Myers, 1982

Anderson Borthwick and Charlie Schroeder stand in the heart of the Park they shepherded into existence

A broad view of the Wild Animal Park's Heart of Africa walking safari

Cheetahs in Heart of Africa with a full view of the Eastern African enclosure beyond

Visitors feeding an Indian rhino during a Wild Animal Park photo caravan

Four of the many Wild Animal Park's breeding successes:
Southern white rhino and baby
Arabian oryx and offspring
Przewalski's horse and colt
California condor

The Heat Was Rising

"Charlie, you can't do that!" you can almost hear Frederick Kunzel's words echoing once again around the board room.

That was certainly the tone of this entry from the Board meeting:

"The minutes of meeting of March 31st were read. Judge Kunzel asked they be corrected to reflect that Trustees Chernoff, Klauber, Kunzel and Olmstead voted against the 'ill-advised' resolution relating to the use of the Elmer C. Otto bequest."

But that didn't deter Charlie.

Early one June morning in 1964, a "special" meeting of the Elmer C. Otto Study Committee was called "for the purpose of hearing the report of the society's legal advisor relating to Stanford Research Institute proposal," and to draw to the attention of the Board of Trustees, the "necessity of initiating an amendment to the Ordinance to allow the use of funds, other than tax revenues, outside Balboa Park."

The heat was rising.

At a special Board meeting a week later, called to meet with the Stanford Research Institute officials, this item made its way into Charlie's minutes with—we can only imagine—Lawrence Klauber's firm admonition:

"Mr. Klauber closed by stating that the Trustees have made no commitment concerning the establishment of any form of unit in the San Pasqual area, and they certainly have not yet approved any form of activity in that area."

"It Will Work!"

"The Board kept saying no," complained Charlie. "So we took on the Stanford Research to do the feasibility study. And lo and behold, they said it would make money—it will work; it won't be a failure!"

At the March 1965 Board meeting, Charlie related the good news. The Stanford study was overwhelmingly positive. After conducting surveys and projecting figures, the study indicated that the San Pasqual "back-country Zoo" could prove 82 percent as popular as the parent Zoo but would not detract attendance from the latter. Then, better than that, the report inspired Charlie and the Board members who favored the park concept to think even bigger.

The Stanford Research group asked some concept-expanding questions:

- Should it be a game farm?
- Should it be a game preserve with limited viewing?
- Should it be a full-blown "natural environment" zoo—
a new approach with viewing from a train?

While the researchers were expressly forbidden to give opinions, the study leaned heavily, if not enthusiastically, for option #3.

"The San Pasqual Preserve would be an entirely different form of entertainment than that offered at the Zoo in Balboa Park," wrote Harry Gillis of the Stanford Research Institute. "It is our considered opinion that a train ride would prove a greater attraction than a so-called safari bus."

The Park's concept was about to take another step in its evolution. As Charlie put it, "We had not thought about the Park in large terms initially, but some on the Board, including Andy Borthwick, said, 'If we're gonna do it, then let's do it all the way. Let's make it pay for itself.'"

Charlie was now in high gear.

That fact must have been intensely obvious to each and every Board member by this time because Lawrence Klauber, at the May 1965 meeting, asked that a letter be read and placed in the official minutes:

> I don't wish to see the Zoo go into debt for $1.75 million dollars to establish this millstone around our necks. I don't favor a course which legally or otherwise will jeopardize our city mill tax. I don't favor any preliminary commitments for land or fencing at San Pasqual, a foot-in-the-door policy, which despite all promises and understandings, will put us on a course from which we cannot retreat.
>
> I think this Board should prefer to sponsor one superlative Zoo rather than two mediocre ones.
>
> Laurence M. Klauber
> Member
> Board of Trustees

What was Charlie's response?

Now he felt he needed to take his stand.

Within days, his written response was given to the Trustees, and it contained a chin-out, heads-up, all-Charlie retort:

"In the matter of the San Pasqual development, we do indeed wish to put our foot in the door, and hopefully we will never be in a position of retreat."

That defiant statement was part of a six-page statement of "Management's Concept" for the San Pasqual development, including costs and a confident Schroeder stance buttressed by the Stanford Research Institute's positive study.

He wrote the following for the record, evoking the Stanford study often, making his case and expressing his beliefs strongly and clearly:

> I believe it essential to consider establishment of a supplement zoo at this time to accommodate San Diego families and visitors in the years immediately ahead.
>
> The London Zoological Society discovered this need many years ago and established the 650-acre Whipsnade Zoo, supplementing their zoo in Regent's Park; and rather than creating a conflict of interest or taking funds away from the smaller zoo in the city, the outlying zoo (forty miles from Regent's Park) is supplementing the city zoo's income! The need for an additional zoo in Paris was demonstrated immediately after the war, and the impressive Vincennes Zoo, in another borough not too far removed, was established. Both are self-sustaining.
>
> The San Diego staff is unanimous in its enthusiasm. Full backing of city government including utilities, has been assured.
>
> Trustees asked for reports on three types of operation: These actually are not separate operations but rather step developments. The conservation farm and game preserves are the forerunners of the natural environment zoo. It would be entirely unique, supplementing—not detracting—from the Zoo.

Then after pages of figures and listings, he ended with this:

> With some assurance, we know we will continue to have a superlative zoo and completely different natural environment zoo without in any manner detracting from the present zoo, or overtaxing staff.
>
> On this occasion, we need your enthusiastic support.
>
> C. R. Schroeder

Into the News

With Charlie's talent for moving public opinion, the fact that newspaper and magazine articles about the proposed animal park began to appear during these months is not surprising. Local columnists and reporters called Charlie continuously looking for Zoo news, and he was never shy about feeding them the latest and juiciest morsels. Clearly, it was time to feed them the whole enchilada.

"An Africa for San Diego?" began one article in the *San Diego Union.* "All aboard—a diesel-powered tractor pulling the canopied passenger cars gives a few low-gear jerks and starts on its six-mile journey through the wilds of Africa. That's the concept of proposed San Diego Zoo's back-country natural habitat zoo."

But the boardroom struggle was also finding its way into print as well, complete with a nicely quotable decision deadline offered by "staff," which, of course, was Charlie Schroeder:

"Land owned by the city, managed by the city department of utilities and has been offered to the zoo for use as a public zoo that would stress conservation and might also lend itself to a breeding area, supplying animals to San Diego and other zoos," the *Union* article announced. A majority of board is for it, the article said, then ended with this interesting quote from Charlie:

> "The opposition," Dr. Schroeder explains, "feels all our energy should be devoted to the Zoo in Balboa park. They feel the Zoo in Balboa Park will fail to grow and prosper if the staff is dividing its affections. Those who favor the back-country zoo cite the findings of a Stanford Research Institute report last year, which says the new

> zoo would be unlikely to detract from San Diego Zoo." Despite obstacles, the Zoo staff talks positively. They say, they will have it eventually. The official announcement could come this year.

Still, nothing he tried worked—no minds were changed, no votes reversed. The stalemate remained.

Interestingly, during this time, one of the Board meeting minutes read: "Old Business: San Pasqual development. Society not in position to lease—permissive right to use discussed."

It's as if Charlie added, *Well, okay, how about if I can get the land without a lease? Could we make the Park happen then?*

You can almost hear Laurence Klauber saying, *Charlie, if you bring this idea up again, I'll vote to have you fired!*

Another year would pass before the mention of the San Pasqual Game Preserve would surface again in the Board's meeting minutes in 1966.

In that year, Charlie would be coping with personal tragedy as his wife, Margaret, died of cancer. The Park and the future would have to be put on hold.

CHAPTER 19

Something Great Very Soon

1966-1967

In the spring of 1966, Zoo-50 had the whole city celebrating the Zoo, and between Charlie and Zoo-50 Chairman Sheldon Campbell, the emphasis was on conservation.

The International Conservation Conference created a stir. Renowned experts from around the globe discussed the problems and solutions Charlie had been espousing locally for years.

Charlie must have gotten an incredible amount of energy from it all, enough for a second wind—or more rightly, a fifth one.

Better yet, a new president of the Board had just been appointed, Eugene Trepte, who had always seen the promise of Charlie's tenacious idea.

In a special *San Diego* magazine Zoo-50 issue, Gene expressed his feelings about Charlie's vision:

> "I probably am more enthusiastic than anyone else about the San Pasqual game preserve," says Gene Trepte,

young president of the Zoo Board of Trustees. "Maybe it will take ten to twenty years. But there is no doubt that with increased population and increased leisure, there is more and more craving for recreation and places to visit.

"Besides, we know that it is going to be harder and harder to import animals from Africa," he went on. "It would help our position in trading to have this preserve, where herds could increase in size. Really, only three things are holding us up on this now: a workable arrangement with the city of San Diego, funds to start the project, and funds with which to operate it until it shows a profit."

A New Big Idea

Ten or twenty years? Charlie didn't have ten or twenty years. He did have a different thought, however—and it was a good one.

First, though, Gene, as if he were raising a finger to the wind, checked with the Zoo's lawyers. He asked them: Has anything changed if we try this again?

Except for the passage of time, they told him, there had been "little, if any, change in connection with this matter since their discussion on the subject a year ago." So the new Board president decided to see if he could be the one to get this "zoo of the future" on its way. In June 1966, the minutes state that the San Pasqual game preserve idea was "reopened by Eugene Trepte."

Carl Hubbs, Charlie's first Board member champion, offered up Charlie's new big idea—a resolution that the president appoint a committee to study the game preserve concept and report to the Board "without making advance commitments."

President Trepte did just that.

Norm Roberts, who was appointed chairman of the new San Pasqual Study Committee remembers it this way: "Charlie convinced them to put together a study committee of non-Zoo Board members, a blue ribbon panel consisting of people from the community."

Charlie once again turned to the community—and not just San Diego. The Wild Animal Park was going to be on the very edge of a town called Escondido. It would be a boon for the mid-size city if its people decided they wanted to be a part of Charlie's dream.

Ribbons and All

J. Ray Baker, president of the Escondido Chamber of Commerce, found himself sitting by Zoo director Dr. Charles Schroeder—no doubt by pure Charlie coincidence—on a group bus trip in 1966, during which Charlie told him of his plan to create a wild animal park in San Pasqual Valley.

"He talked about the great opportunity it would create to preserve animals that were facing extinction," recalled Baker. "He said he was facing a stone wall, however, in the Board's thinking: 'Why risk our Society out in the back country when we have the best zoo in the world right here?' That is how Charlie explained their stance."

Baker asked if the Escondido Chamber of Commerce could help.

You know what Charlie's answer was.

Baker made a pitch to the Escondido Chamber, procured a resolution, took it to the Escondido City Council, received another resolution, ribbons and all, and at Charlie's request, presented both resolutions at a Board meeting. "It would be a recreational and economic windfall," Baker told the Board.

Baker said Charlie appointed him to the Study Committee and years later, when Charlie moved near the park, he and Baker became close friends.

Back to the Hill

During this time, Charlie continued his treks to the hill above that special San Pasqual valley. "He spent a lot of time with the ones against the project, showing them every damned thing he was going to do and how he was going to do it," remembers Gene Trepte. "He forgot the other nine were with him. But Charlie was smart. He said, 'I've got to take them and make them walk where I want the railroad to go and show them why I want to do it this way and that.' That was the unique thing about him. He was superb."

He also pulled artist and designer Chuck Faust into the thick of things again. After all, seeing is believing, and since it had worked for the Children's Zoo, maybe some visuals were in order. But this time, the staff came up with something better than a fistful of sketches. Chuck and his colleagues decided to build a three-dimensional model that would be an African feast for the eyes. By August 1966, this item made the papers:

"Yesterday, Zoo officials announced they are building a three-dimensional mode of the outdoor zoo, which will feature animals from antelopes to elephants as they would be seen in their natural habitats."

Charlie's way with the press had now extended to the *Escondido Times-Advocate,* as this story attests:

> They have the Stanford Report. What then is holding up progress on the San Diego Zoo's San Pasqual Game Preserve? According to Bill Seaton, the Zoo's publicity representative, "Some members of the twelve-member board of directors aren't completely sold on the idea." Directors presumably could have proceeded earlier, as it has been reported that a majority are in favor of the project. "However," said Seaton, "they prefer that the board's opinion be unanimous in this regard."

Discouraging Word

Then, Charlie who had courted the city's reporters for years, was blindsided by a negative editorial that appeared in the *San Diego Union.*

> The San Diego Zoological Society is correct in using a cautious approach to the proposed development of San Pasqual Valley into an African-type wildlife adjunct of the Zoo.
>
> The idea of putting large African animals in a natural setting in San Diego County is immediately appealing and creates the temptation to rush ahead. Economic factors should dampen the urge. Before the additional area is added to the Zoo, the Society must be sure it will be economically autonomous. It also must be certain that the proposed area would in no way detract from the present facilities in Balboa Park. Thorough studies of these and other points by several groups, as approved by the Society, is the proper approach.

Then a letter to the editor from Board member Howard Chernoff found its way into print:

> I have been reading stories and suggestions indicating that the proposed San Pasqual Zoo is almost a fait accompli and that only a few details remain to be solved before the Zoological Society announces a grand opening sometime in 1970. I believe it is time the public has the facts as the Zoological Society's Board of Trustees knows them. I think it might be well for the public to view this project with the picture of the presently successful Balboa Park Zoo in its mind because nothing must be done to harm it.
>
> Howard L. Chernoff
> Past president
> Zoological Society and Board of Trustees member

The plot had thickened, and it was being played out in the press for all to see.

After that—and perhaps because of it—Eugene Trepte, at the first San Pasqual Study committee meeting, gave copies of all previous news releases and stories to the members, then requested everyone to refer reporters and other interested persons to Bill Seaton, assistant public relations director. Seaton, from then on, would be the only source of authorized information on the topic.

Thus ended the dueling newspaper accounts, at least for the moment.

Something Great, Very Soon

Except for his ongoing trips to his favorite San Pasqual hill with old and new friends alike, Charlie must have realized it was up to the study committee now.

President Norm Roberts recalled that the committee weighed the three biggest questions:

Should the park be done?

Where would the money come from?

Was there any hope at all that the Board could unanimously agree if the money was found?

As the months passed, a strong suggestion made by member Oscar Kaplan—that the study committee should present "something great, very soon"—must have made Charlie smile.

Yes, very soon. Surely Charlie agreed. He was sixty-six years old. The retirement age that he helped set at the Zoo was sixty-five, so he knew he was working on borrowed time.

The committee's response had to be right one, and it was.

In mid-1967, the San Pasqual Study committee formulated a formal report, and one statement in the midst of all the words was probably music to Charlie Schroeder's ears:

"A wild animal park is necessary."

The Vote

Now the heat was not only rising, but the water was boiling. Board president Gene Trepte felt it was time. In May he called for a special Board meeting at a time when the greatest number of Trustees could be present for the express purpose of considering the San Pasqual project.

In attendance on June 22, 1967, were Andy Borthwick, Howard Chernoff, Minton Fetter, Gordon Gray, Carl Hubbs, Laurence Klauber, Fred Kunzel, Lester Olmstead, John Scripps, Robert Sullivan, Eugene Trepte, and Milton Wegeforth.

Also in the room were Chuck Shaw and Dr. Charles Schroeder.

From the minutes, this was the scene:

> The meeting was called for the purpose of deciding the Society's participation in the development of a wild animal park in the San Pasqual area of the city of San Diego.
>
> It was moved, seconded and passed that the Trustees approve *in principle* the San Pasqual Wild Animal Park with the proviso that a plan of financing be developed acceptable to the Trustees.
>
> It was moved, seconded and passed that the president appoint a committee of three Trustees to explore methods of financing.

"In principle" were the key words, of course, but it was the first positive vote.

As Norm Roberts remembered it, "They okayed a go-ahead but with the understanding that no Zoo funds would be used and that the way of raising the money be approved by the Trustees."

Then Norm recalled another pivotal moment: "Suddenly Andy Borthwick called and disbanded the study committee. He'd decided to make it happen. Charlie had persuaded him."

Anderson Borthwick—who would follow Trepte as the next Board president—was poised to make the San Pasqual project go.

"We just wore them down," Charlie said, and it was true.

But the fight wasn't over. Finding outside funding was the final hurdle, but Borthwick folded the study committee into a special finance committee and began hammering away at this final obstacle. Charlie, however, had something else on his mind—his up-coming marriage to Maxine Dawson. Meanwhile, the dissension continued on the Board, so much so that their honeymoon even affected the situation, according to this mention in Neil Morgan's column on October 31, 1967:

"Dissension between directors and management of the San Diego Zoo hasn't seemed to take a holiday during Dr. C. Schroeder's wedding trip. It centers around budget affairs and the proposed San Pasqual animal reserve."

Changing of the Guard

Suddenly, there was a changing of the guard on the Board. With the passing of Gordon Gray and Laurence Klauber, and Howard Chernoff's pending resignation, the only staunch dissenter left was Frederick Kunzel.

And don't you know that Charlie took each new Board member straight to his hill above the San Pasqual valley and shared his arms-wide vision just for them.

The dream was within Charlie's sights.

CHAPTER 20

A Rap of the Gavel

1968

Charlie and Andy Borthwick became quite a team.

Anderson Borthwick was a six-foot, six-inch former Olympic rower, and Charlie was a five-foot, six-inch former tumbler. The two were both pushing seventy, and yet they remained formidable personalities. If synergy is defined as something greater than the sum of its parts, then these two were the pure definition of the word.

Victor Krulak remembers their synergy well: "Anderson Borthwick was very strong and he growled at Charlie a lot, but Charlie really was in charge. 'Whatever you say, Charlie,' he'd always answer with a bit of a grin."

Sheldon Campbell, one of the new Board members, described one of these pivotal 1968 meetings this way. "The first Board meeting I attended was to vote whether or not to go ahead with the Park. I voted in favor, but I couldn't see where the finances were coming from—there was no financial plan to pay for the Park. What I didn't know was that Andy Borthwick had plans."

As Charlie and the members of both the Board and the special funding committee knew, Anderson Borthwick was a legend in the San Diego banking community. When Andy said outside funding would come, somehow, some way, it would be there.

The Big Plan

Things began to happen in a hurry. While Eugene Trepte began working on the agreement with the city, Charlie and Andy Borthwick began to work with the banks.

Then, during a funding committee meeting, Norm Roberts suggested a severe answer to funding but one that could work—municipal bonds. If any city would back a wild animal park, pointed out Roberts, it would be this one.

Charlie remembered that Borthwick, on hearing that idea, had second thoughts about getting the banks involved. Before trying more drastic measures for the $6 million needed, "Andy just up and decided we'd go for a general obligation bond." That would take months, however, or longer. Right now, they needed money to secure the lease and to begin improvements such as fencing.

Suddenly, someone—a friend of Charlie's would be my guess—offered a special bequest of $48,000 to the Zoo specifically for the San Pasqual project "to start operations." As for more start-up capital, they devised some workable possibilities. For instance, the new Skyfari was making money hand over fist. As Charlie said, "Skyfari cost us much more than we thought, but we made money twice as fast as we thought."

So, they must have brainstormed, *why not add a little Skyfari money to the $48,000 bequest and a chunk of the Otto money in order to secure the lease of the property and get the Wild Animal Park fenced and going?*

Then, we'll aim for a bond vote to build the Park.

They had a plan, a good one. A great one. It could work—it would work—because it was time.

A Ramrod Rap of the Gavel

On September 5, 1968, funding committee chairman Anderson Borthwick called together the Board members for a breakfast meeting, along with advisors such as Norm Roberts and Victor

Krulak. Andy Borthwick called the special meeting for one purpose, and one purpose only.

The minutes tell the tale of a dramatic opening: "Meeting called to order at 7:40 a.m. by chairman Anderson Borthwick, who briefly summarized the ten-year history of the Wild Animal Park project, then announced proposed plans for funding."

Charlie would enjoy telling this part of his decade-long saga for the rest of his life. One reporter—who heard the original rendition from Dr. Schroeder—wrote: "With obvious glee, Dr. Schroeder recalled how he and Borthwick conspired to win the final vote."

In his last interview before his death, Charlie told the story with the same "obvious glee":

"This is the absolute truth. We held a meeting to pass the operation. We didn't have a quorum, but Andy said, 'Gentlemen, this is why we're here. To vote on the Wild Animal Park.' Then he popped the table with his gavel and said: 'Passed!' "

At this point, Charlie laughed loud and satisfyingly long, then added: "We had finally ramrodded it through."

Finally.

How did it really happen? Probably much like Charlie remembers and how the legend that sprang from those pivotal years tells it.

The minutes of that special breakfast meeting say this:

> The point was made that the Zoological Society funds are presently available, together with a $48,000 bequest made specifically for the San Pasqual project, to start operations; however any funds borrowed from present reserves should be repaid....
>
> It was the CONSENSUS RECOMMENDATION *[capitalization courtesy of Charlie Schroeder's minutes dictation]* that development of the area be initiated as soon as feasible and practical.

At the next regular Board meeting, the special bequest plus the bond idea were announced and shortly thereafter, $500,000 was voted to close the lease and begin building the park.

The plan had worked. Charlie's foot in the door, along with Borthwick's, Trepte's, Roberts', and other faithful Park supporters, had finally kicked that gate wide open.

Celebrating on the Hill

Charlie's dream now was in the voting hands of the people of San Diego. That thought must have given Charlie much comfort and great confidence. He would have to wait two more years before the bond vote, but that wouldn't slow down Charlie Schroeder, not after all the years it took to get here. The Zoo-loving people of San Diego would make the rest of the money appear with a simple vote, all in good time, Charlie must have fervently believed. Until then, with start-up money ready and waiting, there was work to be done, and a long-awaited dream to begin.

Charlie celebrated in the way he knew best. As Neil Morgan's column on November 22, 1968 put it:

> Dr. Charles Schroeder, the Zoo director, ascended a rocky knoll overlooking the San Pasqual Valley this week to describe plans for the Zoo's proposed wild animal reserve east of Escondido.
>
> His sermon on the mount, before two busloads of San Diego businessmen, told of a dramatic 1,800-acre fenced Zoo Annex through which visitors would travel by train, seeing greater concentrations of wild animals than on many East African safaris. With good fortune in city and community cooperation and fund-raising, the reserve could be ready within three to four years.

First Spade of Earth

When the final lease agreement was reached between the city and the Zoological Society, the official ground-breaking ceremony took place on May 14, 1969, ten years from the time that Charlie and his staff had begun to brainstorm what they should do with their population surplus.

While Miss Zoofari, a guanaco, a baby elephant, and a beaming Dr. Schroeder stood by, Anderson Borthwick and Escondido Mayor William Crow turned over the first spade of earth on Charlie Schroeder's peaceful animal kingdom to come.

CHAPTER 21

—Building the Park—

Out of the Desert Dirt

Charlie was wasting no time now. This part was going to be fun, and he wasn't waiting a minute longer.

In fact, by all accounts, he celebrated by walking the entire perimeter to get the feel of the Park. "I don't think anyone's done it since—lots of rattlesnakes," he said.

This article appeared in the *San Diego Union* in February 1969:

> Work on the first zoo of its type is expected to begin in ten days in the San Pasqual Valley. The area will be fenced in such a way that visitors can safely see the animals as they exist in their natural habitats. Plans also call for a small railroad to run around and through the sections of the zoo. More than ten years of research on the wildlife preserve have been done by the Zoo's staff, and Schroeder has been conducting information tours of the San Pasqual site for several years."

Charlie's dream had been called many things: the Back-Country Zoo, the Zoo Extension, the San Pasqual Game Preserve, the Wild Animal Park at San Pasqual—even Zoolandia and SAFARI were suggested. At a special two-day retreat, the Board offered five names:

- San Diego Wild Animal Park
- San Diego Wild Animal Land
- AnimaLand
- San Diego Safari Land
- San Diego Wild Animal Safari
- San Diego Wildlife Park

They finally settled on one name. The Zoological Society of San Diego's zoo of the future would be named the San Diego Wild Animal Park.

Charlie was past retirement age—far past it now, actually. He was pushing seventy, but he wasn't ready to retire yet. "I was just warming up at sixty-five," he told a reporter, and he wanted to see the Wild Animal Park through to completion, as did Andy Borthwick. Both of them would hold on beyond the usual retirement age to see the Park become a reality. Denver Zoo director Clayton Freiheit remembers all the years Charlie shared his Wild Animal Park dream with others in the zoo world:

"He told his zoo colleagues all about it," said Freiheit. "He'd bring it up at the meetings of the AAZPA (American Association of Zoo, Parks, and Aquariums), and most of us were incredibly envious. He and Andy Borthwick finally made it happen, and the San Diego Zoological Society is certainly the better for it."

In a 1970 newspaper article about the pair, Charlie and Andy, they were described richly:

> In this, its greatest expansion year, the San Diego Zoological Society is led by a pair of energetic septuagenarians. Most American organizations decree retirement at age sixty-five. Wisely, the Society does not. Its president Anderson Borthwick, seventy-two, and its zoo director, Dr. Schroeder, seventy, can outwork and outthink many men who are younger calendar-wise by a decade or more.

Picnic on His Hill

Charlie organized his first planning picnic. As he explained it, "Oscar Kaplan, chairman of the P.R. committee, Robert Jarboe, superintendent of construction before Hal Barr, the city manager and myself—we took Colonel Sanders and beer, went up on the hill, and dreamed it!"

Of course, the first practical action was to fence the perimeter of the property, which would be no small chore. Much of the material had to be brought in by helicopter because of the rugged terrain with no road access.

"We thought we'd put in a peripheral fence then build separated inside fenced areas for various exhibits we had in mind," explained Charlie. "I say 'had in mind,' because, let's face it, there was no set formal plan of what we'd use and how we'd use it. It sort of grew as we went along."

Charlie began to hand out the plum assignments to the lucky pioneers.

Everyone said yes, of course. All the construction went to Hal Barr; all the animal purchases to assistant curator Jim Dolan; all the medical practice to Lester Nelson; and all the design to Chuck Faust. "They all took part in deciding the size of the enclosures and enclosure techniques, whether we would dig for moats, put up stone walls, put up fencing, in-riggers, the whole thing," as Charlie explained.

Ian Player's First Sight

Ian Player and Nick Steele, both of Zululand, South Africa, visited during this time. Player was chief conservationist for Zululand and renowned for saving the Southern white rhino from extinction.

"You have a wonderful opportunity here to make a great contribution to the conservation of wildlife," he told the Zoo group as he overlooked the future Park. He gave technical advice on type and height of fences, noting that in some game preserves in South Africa they had witnessed kudu antelope jumping a six-foot fence with no effort at all. He also made suggestions on the size of herds to be placed in the two principal valleys. Ian Player only had one caution: "Put up fences inside, and you'll ruin this site."

This wasn't Ian's first look at the San Pasqual. Back in 1964, he was one of the first people Charlie took to his site on the hill. Ian must have looked around and seen a grand place for his rhinos. He told Charlie the place looked like home, and he thought the Southern white rhinos could thrive and reproduce here.

Soon, that thought would turn into a story that would become a Wild Animal Park legend.

Who Are You?

Ian remembers the first time he ever met Charlie Schroeder. He had come to San Diego as a guest of the Metro Goldwyn Mayer movie studios. "They'd made a movie called *Rhino,* and I was traveling all over America advertising the movie in 1964," he said. "I'd never been very keen on zoos, but I realized because of the situation in South Africa at that time, it was important to have rhinos in zoos as an added protection for the potential gene pool."

Ian visited many of the major zoos around the country, finally stopping by San Diego for the specific purpose of being filmed in front of the Zoo's two Southern white rhinos. "I remember very clearly walking in the main gate and seeing the flamingos. I was enormously impressed with all that I saw," he said. "There was a huge difference between the San Diego Zoo and every other zoo I had visited. I immediately knew I was in an efficient and well-run place."

His MGM agent had called the San Diego Zoo to set up the filming in front of the rhinos. "That's when Charlie Schroeder came on the scene," remembered Ian. "He didn't care if it was MGM or Warner Brothers calling, he wanted to know who we were because this was his Zoo. I remember hearing the agent stumbling and muttering some answer to him. Schroeder said he'd think about it and make some inquiries. I was relatively well-known in the animal world at that time, but Schroeder would not let me in front of the enclosure until he had checked me out, and I was delighted. It showed me he was in charge and he wasn't going to be overruled by anyone. I could *see* that the Zoo was good, and now I *knew* it was good because of this man."

Ian remembers telling Charlie that the rhino enclosure was quite small, not to mention that it was all concrete. He added that the rhinos looked perhaps a bit too well-fed and that the Zoo might not be able to get them to breed. "That's when, way back then, Charlie mentioned the possibility of the land being available for a Wild Animal Park. He asked me what I thought, and I replied that it was an excellent idea." Charlie took Ian and probably a bucket of chicken to his favorite hill and preached what Neil Morgan would later call his "sermon on the mount" to this South African.

Then it was Ian's turn to capture Charlie's imagination. "I suggested that they split up the exhibits into continental exhibits and put in water holes. I had very recently visited Lion Country Safari Parks in which cars drive through the actual herds, so I said automobiles were a no-no. I was adamant about it. That was the birth of the monorail idea."

Charlie must have stored Player's ideas and suggestions away, saving and savoring each one of them, because ultimately they all became integral parts of the Park's development.

Bond Vote

"Now we had to get the money," as Charlie put it. "A lot of it."

The bond vote had been set, which prompted this coverage from the *San Diego Union:*

> San Pasqual Valley could easily be called Kenya without zebras. The scenic valley about thirty miles northeast of San Diego closely resembles the rolling terrain of the East African nation, which is synonymous with safari. In the next five years, San Pasqual Valley will become Little Kenya as the zebras become part of the landscape along with other African animals such as antelopes, elephants, lions and ground birds. It will be the only one of its kind in the nation. The whole concept is to create a water-hole environment.
>
> "It's unique," said Dr. Schroeder, the nearest thing to game preserves of Nairobi, Kenya. "If I could have the privilege of going almost anywhere, I don't think I would have picked another site for this zoo."

Then the *Union* come out editorially in favor of the municipal bond vote. "The San Diego Zoological Society deserves wholehearted support in its farsighted, challenging project of creating the

Wild Animal Park," stated one of the newpaper's editorials. "This approach to animal displays will crystallize further the Zoo's position of world leadership in the departure from the traditional bars-and-cage exhibits. The San Pasqual plains will be a further step in natural habitat display of animals."

Wild Animal Park Bond

Positive news articles and editorials would not be enough to assure a winning bond vote, however. As Charlie told more than one reporter, "Community help will be a central key to this project." That was Charlie's theme with every such opportunity because, as the bond vote loomed near, community help was exactly what was needed. Since no bond issue had passed in San Diego for years, Charlie, Andy Borthwick and the Wild Animal Park Committee decided to take no chances. They made plans for an enormous campaign.

Charlie and finance committee chairman Ivor DeKirby asked Dr. Albert Anderson, a well-known philanthropic dentist and personal advisor to Mayor Pete Wilson, to head a citizens' committee to make it happen.

New Board member and publicity whiz Bob Smith helped produce a film called *And Then Came Man* with the participation of Meredith Wilson, the creator of the musical hit, *The Music Man.*

Kids were used to canvas door to door. Every organization and club willing to listen was visited by Zoo people, from keepers to curators to directors to committee members. The city was blitzed, and everyone got into the act. KFMB, a local television station, aired an editorial that began this way:

> It's not often that voters have an opportunity to approve a bond issue that is not a burden on the taxpayer. This is a $6 million bond issue to finance the first phase of construction of the San Diego Wild Animal Park in San Pasqual Valley. This park will be owned by all the citizens of San Diego, and every cent of the bond issue is to be repaid by the Zoo. Even under the worst circumstances imaginable, such as the complete loss of revenue at the Zoo, the owner of a $25,000 home, for example, would be

taxed only $2.25 per year. Proposition B is a bargain that San Diegans cannot ignore. Vote "Yes" on Proposition B.

The Proposition

PROPOSITION B. City of San Diego Wild Animal Park Recreational and Educational Facilities Bond Proposal: to improve, develop and expand the area of the San Pasqual Valley known as the San Diego Wild Animal Park, shall the City incur a bonded indebtedness in the principal amount of Six Million Dollars to permit the acquisition, construction and completion of facilities to provide recreational, educational, scientific, ecological and research facilities in harmony with the open space concept of the valley?

As it turned out, the $6 million bond issue on the November 1970 ballot passed in a landslide with 75.9 percent of the vote. All the other bond issues were defeated that day, but the one for the Park stands today as the largest "yes" margin in city history.

In Charlie's files, a handwritten letter was found from the month before the vote. It must have touched him. In the cramped, earnest style of a teenager's hand, the missive said:

Dear Sirs:

I am Mr. and Mrs. Griffith's sixteen-year-old daughter and am interested in helping getting Proposition B passed. I would rather help in some other way than going door to door, but if that is all I can do, I will do that in grander style. I could stand in front of our supermarket and talk to the shoppers. Thank you dearly.

Sincerely,
Linda Griffith

Can you imagine Charlie's excitement? This was the final hurdle, a leap of faith offered by his beloved city.

In the next two years, he was a one-man dynamo in transforming the hills and valleys into a world-class animal park.

Team Effort

Charlie was the first one to say that his vision for the Wild Animal Park was the summation of a Herculean team effort, from citizens' committees to teenage canvassers to staff to people on the Board, especially Andy Borthwick.

Of course, that was a sincere yet charmingly calculated way to get everyone working to make the Park "our Park," just as the Zoo had always been "our Zoo."

There was even a fun citywide competition to see who would donate the first animal to the Park, as seen in Neil Morgan's *Tribune* column from February 1969:

> It may be a photo finish in the race to be first to contribute an animal to the Zoo's San Pasqual Wild Animal Park. The Boy Scouts were told that they could be first and announced this week they would buy a buffalo. But Southern California First National Bank got there first with the money for three East African antelopes in honor of Anderson Borthwick. The Scouts meanwhile have decided they won't buy a buffalo after all (not glamorous enough), but a silver antelope.

That must have inspired Charlie, because by May, a newspaper article quoted the cost of an Angolan giraffe ($3,500), a rare white rhino ($10,000), and thirty ostrich chicks from northern Massai (also $10,000). "We have published these lists of animals depicting the hundreds of animals needed for the Park," Charlie told the reporter. "Some of the animals to be shown at the Park will come from the Zoo, which is overcrowded with some animals right now. But a great many others will come from organizations, individuals, and groups in the county. It would be exciting for a club to donate an animal or equipment for a project like this," he hinted in print. "Community help will be a central key to this project. Who knows? Perhaps somebody will even donate a train."

Charlie didn't forget the trees and plants, either. "Planting will begin by summer," he added, "as soon as the water wagon arrives. We have 50,000 trees and shrubs ready for transfer now, and we are looking for more. Anyone in the county who has palm trees or other plants can call the Zoo right now. We will consider it a donation and send someone out to pick it up."

That's the way the Park began. By the time the Park opened in 1972, Charlie would be quoted again on the subject in the *Escondido Times-Advocate* in a special supplement commemorating the opening of the Park. School children in the county had contributed about $9,000 in pennies to buy animals—mostly lions, he said. "The Rotary Club of Nairobi has sent pictures and information to guide us in designing and building Nairobi Village—they were delighted when we chose the name of their city. Margaret Kenuyatta, mayor of Nairobi, has written to us personally." Campfire Girls presented a check for a pair of secretary birds. Teachers worked with the Zoo's education committee to brainstorm ways for the Park to be a learning tool for San Diego children. The list could go on for pages, and Charlie would have delighted in every word.

"The most exciting thing about this project is the fantastic enthusiasm and support it has drawn from all sorts of people," he told a newspaper reporter.

Shingles, Zebras, and Mules

The same excitement flowed through the construction of the Park as well. Since the major construction funding was coming from a general obligation bond, the city had to be involved in all construction planning. After ground broke, however, lumber and shipment strikes suddenly happened across the state, and the large contractors said there was no way they could get the Park built on time.

Charlie and the Wild Animal Park committee didn't know what to do until Charlie talked to the Lou Pauletto Construction Company, an Escondido firm. As Charlie would say later, Lou was so excited about being a part of such a big challenge that Lou would claim that he didn't know about any of those strike "problems."

He got the job and had a great time. All the shingles, for instance, were curved because Lou devised a special scheme of steaming and bending them. "He worked just from Chuck Faust's sketches—not specific working drawings," said Charlie. "When Lou was finished, he said to me, 'What will I do now for a challenge?' He finished two-thirds ahead of schedule, and he gave all his employees memberships to the Park."

It took longer than expected to finish erecting the perimeter fencing, but since animals were already arriving, holding areas had to be quickly constructed. "We were buying twenty Grevy's zebras at a time," said Jim Dolan. "We were importing in big lots, which isn't done today." Soon the fences were up, the holding pens were opened, and the animals began to roam around their new, wide-open home.

Two park rangers were hired to guard the new Park.

They began to patrol the fence line, noting any breaks or evidence of predators. There were miles of perimeter fencing to patrol, some on very steep terrain. A sure-footed mule named Patrick, who held staring contests with the strange creatures inside the fences, was the answer for the steepest areas. Part of the Park's lore was the way Patrick, safely separated by moats, would stop short, almost losing his rider, to peer for minutes at a time at the rhinos, zebras, or whatever creature might be staring back, before he deigned to turn and go on his way.

Designing the Park

For the man who had helped transform the Zoo into a moated trendsetter, Chuck Faust must have felt that lightning had struck again. That could be enough creative accomplishment for one life. Now he was being asked to conceptualize a zoo almost without boundaries that would become the envy of the zoological world.

The Board had decided to build a village at the Park entrance with smaller but still spacious exhibits, where visitors could buy refreshments and stroll near a few small, moated animal enclosures while waiting for the Wgasa monorail at Simba Station.

They chose architect Frederick Leibhart to work with Chuck Faust and his drawings. Leibhart's firm gave Faust a desk and got to work.

Faust first thought was to create a replica of old Nairobi.

"To design a village in an African motif when I wasn't sure if such a motif existed meant I had to study all the books and then go there," said Chuck.

"We sent Chuck to Africa with a camera and a sketch pad," Charlie said. "He visited Uganda, Tanzania, and South Africa, where he found thatched huts and burial tombs worthy of his consideration, but he also became aware of a certain style and grace."

"I discovered that the architectural style of the different villages was remarkably similar, even with tribes completely isolated from each other," said Chuck. The entry to the Wild Animal Park is about the same size and shape as a Ugandan king's burial tomb. Faust ultimately would create more than 200 drawings for his conception of the Park's village.

"We call it Faust's Africa," quipped Andy Borthwick.

It's also not surprising that with all this space, Chuck Faust decided to reproduce one of his first big successes. An enormous flight cage measuring twice the floor space of either Zoo aviaries was designed for the entry area.

Of course, some of his more "authentic" ideas didn't make it off the drawing board. "We wanted to use real savanna grass for the roof, but the building department said no in spite of our sending a batch of grass to Los Angeles to have it fireproofed," said Faust. "When the South African government heard what we proposed to do, they offered to send a master thatcher and helper. We wound up using thatch for utility structures and offset shingles for the main building, which turned out looking pretty good after all." That's when Lou Pauletto and his construction firm came up with his ingenious answer of curving the shingles by soaking and forming wood shingles in a curved press of his own design.

A Builder's Dream

Meanwhile Hal Barr, the Zoo's construction chief, was itching to start moving the dirt around. "It's a building contractor's ideal, an enormous piece of absolutely open land that needed everything—utility lines, roads, parking lots, and buildings. This job is the kind of thing a builder dreams of, the kind that only comes along once in a lifetime," he said. "I felt terribly excited and privileged to have a part in creating this park."

Hal, of course was inspired by many meetings with Charlie and his Park-dreaming staff. "We want people to feel they have left civilization behind in the parking lot," Barr explained to a reporter. "By the time they're in the village, they should have shed their frustrations and become completely relaxed part in another world, a world where man is at terms with nature and where noise and pollution are unknown."

More Than Two by Two by Two

As the bulldozers began to move tons of dirt, the animals began to arrive. Charlie gave Jim Dolan a free hand to gather the animal collection at the Wild Animal Park for good reason, and it wasn't just because he came from the wild streets of New York, where Charlie grew up. Dolan was the Zoo's assistant curator of birds who also had a deep, scholarly expertise in hoofed animals. Charlie knew he had his man.

The lucky Zoo animals picked to live at the Park began moving in by late 1969, just as soon as holding pens and some fencing could be built. By the time double fencing was finished, the Park would have twenty miles of wire fencing—eight feet high for exterior fencing and six feet high for the interior fencing.

By the following spring, giraffes were on the plain. The giraffes came, not two by two, but three in a very big truck. As a March 1970 *Union* article about their journey began: "Add Bingo, Blackjack and Ginger, three young giraffes, to the population explosion at the San Diego Wild Animal Park."

By the end of that year, about 200 animals were on the grounds, including wallabies, cheetahs, zebras, waterbucks, and storks.

Jim Dolan gave priority to endangered species and "those that excite the public" in making up his acquisition lists, while he worried about composition problems. Some animals simply couldn't be put in with other animals, obviously, because they'd kill each other. With others, it was just guesswork whether they would harm each other. Yes, they wanted the Park to simulate the wild, to a point. Zebras live in the wild with antelopes, for instance, but do they harm them? Jim found out they do.

"We didn't expect it, so we just removed them," he said. "One of the key goals of the park is the establishment and maintenance of an ecological balance between the number of animals and resources such as space and plant life," said Jim. "That would take time and good, educated guesswork, then a lot of watching and waiting to see if the groupings worked."

Jim was also in for some nice surprises during his shopping spree. He thought he would have a problem finding some of the coveted hoofstock due to a Federal Department of Agriculture restriction that ruled all hoofed mammals must be born in this country. Then he discovered how many species were available in small American zoos. "There were fantastic collections in Texas alone," he said. Other zoos were more than happy to sell or exchange, even when he was trying to get the only ones available.

Reproduction Experiment Begins

The double fence was up, and the animals were in a habitat as close to natural as possible in captivity. There was no reason why they couldn't start the great reproduction experiment while the Park was being built. In fact, the Park's reproduction potential became quite dramatic when a pair of Northern white rhino arrived from the St. Louis Zoo and another pair from Washington. The Wild Animal Park suddenly had the total population in the Western Hemisphere. Hundreds of birds were being transferred from the Zoo into the giant aviary being built behind the front gates of the Park. Some of the most interesting early mammal additions were eight Przewalski's horses (the second largest concentration of the rare Gobi breed in the country behind Catskill Game Farm in New York): six female African elephants from the Rhodesia Department of Parks and Wildlife Management; a twenty-member crash of Southern white rhinos from Zululand and escorted back on a freighter by Wild Animal Park personnel; and six Arabian oryx, the rarest animal to join the Society's collection at the time.

In November 1972, the Wild Animal Park was selected for the rare oryx "because of its proven success in breeding species that haven't done well in conventional zoo settings," said a news report. The Fauna Preservation Society, the World Wildlife Fund, the London and Arizona Zoological Societies, and the Shikar-Safari Club had pledged themselves to save the Arabian oryx from extinction, and now the San Diego Zoological Society had joined them. As the article stated: "Out of fifty-four species and subspecies with reproduction potential at the park, thirty-five have already successfully reproduced."

By the time the Arabian oryx arrived in late 1972, the Park had only been officially opened for less than six months—since May of that same year—yet already, the zoo world had noticed that the great reproduction experiment was working.

Space Surprise

Some of the Zoo animals who moved to the wide-open space of the Park became wonderful examples of how open space can help animals do what nature intended. Mandhla, a Southern white rhinoceros, lived for nine years at the Zoo without showing any romantic interest in his mate. In 1971, he was moved to the Wild Animal Park with Ian Player's group of Southern whites that arrived from South Africa. Suddenly he was *very* interested. He began to make friends with every female in sight and over the next decade, he would become prolific, producing fifty-six offspring. What was the difference?

"All he wanted was some open space, a herd environment, and more than one female to choose from," explained Jim Dolan.

Mammal curator Randy Rieches was asked the same question by actor Alan Alda, host of a PBS special on the country's "new" zoos. Randy summed up the difference in one word: "Space."

Water Lifeline

Before there can be life there must be water. Charlie knew all along the pivotal role water played in the existence of a wild animal park carved out of high-desert scrub land.

"I want to show you the most important thing at the Park," Charlie had said to me on my first day of work there, and then led me to the Park's water line and water tank.

Establishing that water line was an all-important task for the Park. They already knew they couldn't get all the water the Park would need from a well, but the nearest city reservoir was five miles away. That left the neighboring city of Escondido as their source, but a state law prohibited one community from taking water away from another, so by special dispensation, San Diego had to seek permission from Escondido to pipe their water to the city's new park. Again, Charlie's network helped. The deal went through.

Charlie and his wife, Maxine, built a house on top of a hill not too far from the Park. Charlie would call home and ask Maxine to look down in the valley and tell him how much water pipe had been laid down that day. He was as mesmerized at the logistics of bringing the water lifeline into the Park, as he was about everything around him. "They laid down a sixteen-inch line from Escondido's Reed Reservoir, snaked it down a north canyon, and some of the pipe had an actual bend in it to accommodate the ground," he said. "Some places we had to cut through a hill and drill over twenty feet down. Then they switched to an eight-inch line for the rest of the way to the Park, where we built our quarter-million-gallon tank right on the hill. The original budget for water was $20,000, but it only cost $750,000," quipped Charlie. "That was almost time to get rid of Schroeder."

Garden in a Desert

Armed with water, a desert miracle could happen. "Someday there will be an 1,800-acre garden and forest environment at the Wild Animal Park that will also be a source of replenishing the gardens of the world with plants it almost lost forever." The was the announcement of Ernest Chew, the Zoo's head horticulturist given the incredible challenge of making a garden out of the San Pasqual dust and cactus-strewn land. Water line or no, some fancy water reclamation had to happen.

"The sanitary engineers that designed the water recovery system used the standard formula for a park or any place else," Charlie explained, "but they forgot that practically all the human waste up here is liquid, so they had to redesign and rebuild the unit to accommodate all that fluid. The recovered water was used for irrigation, which meant there wasn't a drop of water that left the park," said Charlie rather proudly.

After that bit of reclamation magic, nothing was going to stop the resident horticulture magicians.

"We are going to make the Park as African as we can," Chew told a reporter. Plans were to make a twenty-five acre jungle—a greenbelt from the boundary of the East African compound to the gorilla area behind Nairobi Village at the entrance of the Park. Chew wanted to plant natural vegetation to balance the animal population throughout the Park. He and his assistants worked out a general plan based on the micro-climates of the area and went to work.

The result became a registered botanical garden that even includes an Australian rain forest area. The contrast of "then" and "now" photos are awe-inspiring. The horticulturists' work is the incredible unsung success story of the San Diego Wild Animal Park.

The Monorail

Next question. How do we get the visitors around this enormous zoo? Would it be steam engine? Locomotive? Electric?

The Board had put a green light on a non-polluting, quiet, economic electric train or monorail.

Charlie took Hal Barr and Chuck Faust with him to continue laying out the monorail design, to finish what he started that day when Zoo photographer Ron Gordon Garrison captured the Zoo's director in full Charlie Schroeder-stride, impatiently pounding out a few stakes himself. According to Faust, seventy-year-old Charlie left the younger men in his dust yet again. "Just walking the monorail line wasn't enough," said Faust. "Charlie kept walking up to survey the stakes from above. He wore us out."

Charlie obviously loved every minute of it. "Hal, Chuck, and I laid the railroad where we wanted it to go," Charlie said of that day. "We carefully eyed it, studied it, and spoke to the veterinarians about the spaces. Hal told me where to put the stakes, I'd whack it in, and Chuck would tie a red ribbon on it. I lost a pair of eyeglasses out there somewhere. Anyway, when we were through and brought engineers in to see what they thought, we weren't but a few feet apart from what they thought was possible."

The monorail is unique for many reasons. Chuck Faust designed the superstructure, including the fiberglass seats, vinyl floor, stainless rails, aluminum framing, and ceramic-coated aluminum ceiling. Each car had space for sixty people.

Charlie was proud of it.

"We could carry people around a five-mile track for less than a dollar in electricity, and it was almost foolproof maintenance-wise," he explained. "It was the equivalent of having cars on balloons."

But the monorail almost didn't happen; even after it was built.

As Charlie explained it, a company named Universal Design from New Jersey forgot to factor in a profit for themselves and went belly up. "They sold it to us for $40,000 a car. Westinghouse Airbrake took over the production for $80,000 a car. Then Westinghouse folded," Charlie explained. "Rohr then took it over—the group who'd just finished the San Francisco BART transit system." The rails themselves were even extraordinary. A common steel rail would rust and flake and that would mean trouble, so the company wanted Charlie to buy Cortan, a special steel with a red surface that stopped oxidation. "We couldn't afford it," said Charlie, "so they *gave* it to us."

Even Rohr, however, would soon be out of the picture, which meant the Zoo would be on its own with its unique monorail. Extra motors had to be stockpiled, and people had to be trained to maintain the train, as well as constantly fight the brush that grew on the electric third rail. Many riders from the first few monorail trips told stories of being stranded with a herd of rhinos between them and civilization.

But it was all worth it, of course. The bugs were eventually worked out, and the monorail is still in use today, going strong. Today's monorail rider still has the same unique experience that the *Los Angeles Times* wrote glowingly about: "While riding the Wgasa Bush Line, the average man and woman on the train doesn't realize that it is a condensed trip through Africa and Asia. To make a similar trip would cost them at least $50,000 and months or even years just to see what you can enjoy here in a fifty-minute train ride."

Wgasa . . . Zulu for What?

And what about that name? The story has become one of those eternal anecdotes that keeps popping up in newspapers and magazine accounts and even on the Internet, year after year after year.

The story is funny and a little unbelievable, but it's true. It started out as a joke after another long, tiring meeting at which Chuck Faust, Chuck Bieler, and the usual gang were sitting with their clipboards and pencils, attempting to come up with a name for the monorail. They weren't having much luck. Everyone was tired and punchy.

Finally, as Chuck Faust tells it: "We were trying to decide on the name for the monorail. I scribbled down 'WGASA' on some plans. Everybody laughed because they knew what it stood for, but they loved it because it sounded African. We thought WGASA would blow over, but it actually stuck."

What did it stand for? Faust's scrawled letters were a popular acronym at the time that stood for 'Who Gives a S—- Anyway!'"

Dr. Schroeder was in Czechoslovakia attending an international zoo conference. By the time he returned, the name had spread so fast that reporters were actually calling about it. Chuck Bieler answered the calls, and ever resourceful, he told them that it meant the "World's Greatest Animal Show Anywhere." That is still the best official answer.

But when that answer was tried on the director, Dr. Schroeder wasn't buying. With only a few days left before the opening of the Park, it was too late to change it, so he called in Chuck Faust and said, "Tell me exactly what 'Wgasa' means."

Faust was sure he was going to be handed his walking papers if he told the truth, but he did. "I told him exactly what it meant," Faust recalled, "and all he said was, 'I thought it was something like that.'"

At the dedication of the Park, trustee Sheldon Campbell told South African Ian Player about the story, so Ian had some fun. When an unsuspecting Board member told Ian the name of the monorail, Player responded with: "What did you say the name was?"

"Wgasa,"the Board member repeated.

"Did you know that is a Zulu word?" Ian said with a straight face.

"It *is?*" the Board member said, surprised.

"Yes," Ian said very emphatically.

"Really! What does it mean?" he asked.

Ian smiled. "Who gives a s— anyway!"

And another legend was born.

CHAPTER 22

Keeper Cowboys

As soon as there was a fence and animals to put behind the fence, there had to be keepers. Wild Animal Park keepers had to be unusual people, "cowboys" really, because no one had actually ever done what they were going to have to do. It was one thing to try to manage an animal in a twenty-by-twenty enclosure. It was another to do so on hundreds and hundreds of free-roaming acres.

Ricky Cuzzone, chief game keeper, moved out to the Park in time for the first load of animals. An *Evening Tribune* article explained well the shock for the animals as well as the challenge for the keeper: "A zebra at the new San Pasqual Wild Animal Park was allowed to go free from a holding pen. The animal looked around the vast acreage a few minutes, then went right back into its holding pen. Officials at the San Diego Zoo's park said the zebra's act wasn't too unusual. Freedom for wild animals that have been enclosed takes getting used to."

There were so many questions: how and what to feed them, what effect the natural vegetation would have on their health and how the open space affected their breeding. Veterinarian Les Nelson once sent a man from his office, outfitted with a knapsack, to walk the entire acreage, probably the only other man who has ever done it besides Charlie, to analyze the rodent population for parasites that might be dangerous to the animals.

In the years to come, the concept would become far more complex than just feeding the animals and hoping they'd reproduce. "There's so much science that goes into keeping these species going," explained Mike Mace, the Wild Animal Park's curator of birds. "Diet alone is a science we have to work out. Hornbills, for instance, eat figs, yet there are forty species of figs in their original habitat. In this country, we have one fig, commercially grown, and one that hornbills have probably never seen before. We decided to grow the exotic trees or plants ourselves to keep the birds and animals healthy so they would reproduce and continue to exist. We've created some firsts in these areas because we had to."

Bunch of Cowboys

"What we experienced was a once-in-a-generation experience," said the Zoo's curator of mammals Carmi Penny, one of the early keepers at the Park. "We broke the rules as we wrote them."

"There was no one to go to," said John Fairfield, another one of the Park's first keepers. "No work like this had ever been done before because essentially the Park is a research project of compatible animal groupings. In a small conventional zoo, where only singular species are housed together, most of these problems had never been encountered before."

The Park's curator of mammals, Randy Rieches, remembered his first days as a keeper much the same way. "It was a chance to start all over. We didn't already have preconceived notions. We were just working on Charlie's dream. We looked at books about how things should be, but the problem was the animals didn't read the books. There was nothing written on placing species together. You can't just throw open crates and let all the animals out and expect everything to work. But we were doing stuff that nobody had the chance to do and we had a ball," Randy admitted.

For instance, before the Park opened, the fence was all drive posts and chain link. The keepers were sure the fencing would never hold the animals. But they found out that the animals did not test it.

"We were building a barless zoo. Even though you've got no bars, the perimeter is sort of like the end of a game trail in Africa. There, the herds go so far and they stop, as if they are at the end of their migration," said John Fairfield. "It turned out to be somewhat the same for these animals."

The lure for the new keepers hired during Park construction was the unknown of wild animals on the loose, roaming yet restricted and the incredible challenges that offered. They were pioneers themselves, zoo cowboys "riding" a range full of very unusual, expensive, and prized animals.

"Any keeper at the Wild Animal Park lived and breathed the pioneer spirit personally—where you have to guard something one minute, feed it the next minute, then wrestle it to the ground the next because you were the only one around to do it," declared Randy Rieches.

Keepers were having to solve strange problems everyday. A good example was the mystery of how the giraffes were stealing the kudus' food from the antelope's special food troughs. They watched from a distance and to their amazement, they noticed that the giraffes would actually lay on their sides and snake their long necks into the kudu feeding enclosures. A giraffe-proof feeding station had to be invented, along with a tree-high feeding trough for the giraffes, which turned out to be feed bags hanging from trees.

John Fairfield, who came to the park as assistant lead keeper, was there for only three years, but those were the pure cowboy years. Hand-picked by Charlie Schroeder, Fairfield had experienced great success with captive breeding of giraffes in Colorado Springs, Colorado. Dr. Schroeder heard about his successes, and although he'd never met Fairfield, he picked up the phone and introduced himself.

"John, are you up for an adventure?"

Fairfield had deep roots in Colorado and had never visited San Diego. Of course, Charlie wasn't going to let that stop him. "John, you've got to understand, we've got cheetahs from southwest Africa. We're going to go to Africa and get white rhinos."

Charlie took John on a tour of the new zoo park. "At that time it was still a figment of an old man's imagination," John recalled. As Charlie walked and shared his vision, arms waving excitedly, Fairfield could see workers putting up the fencing in the hills beyond Whiskey Springs. "There was absolutely nothing there but broken beer bottles and open range," said Fairfield. "There were great big ravines, great big rocks, and a beat-up two-lane road. That was it."

John had brought two pronghorn deer from Colorado. When he released them onto the open acreage, Charlie said to him, "Well, what you think?"

John Fairfield stared at the rocks and the brush and the fencing being put up, and he thought, *Well, this is different.* "I was a cowboy to start with anyway, so I told Charlie to sign me up."

As for memories of Charlie, he recalled fondly that "a working day for Dr. Schroeder was from the time he got up until he dropped dead. He knew I was the same way."

John and the twenty-three keepers he would soon hire were about to make zoological history with almost everything they did at the Park. As he explained it to a reporter in 1970, "We are training most of our keepers on the job, which is fitting for this place. They could not make use of what they had learned in other zoos anyway. In no other zoo, for instance, must the keepers figure ways to keep natural predators away from the rare and costly animals."

Original Tenants

The keepers John hired were chosen mostly by whether they could keep from getting themselves killed working in their wide-open world.

"Our headquarters when we started was about fifty-by-thirty with no water, only three walls and a roof—nothing to even slow the wind," he remembered.

Plus, there were the former residents to handle. Coyotes were digging underneath the double fence, and bobcats were finding their way into the cheap eats, too. The keepers were having to fight daily to save the new, free-ranging hoofed population from the predators. Rattlesnakes were everywhere. With all the earth being moved around, the keepers and the workers stumbled onto far too many rattlers for comfort. They collected more than 200 alive.

Then there were the glorious golden eagles, as rare and special as any of the new inhabitants. They were a problem for the population and for themselves.

The problem, of course, was that the new population of birds were "pinioned"—their wings were clipped. There were hundreds of very expensive birds, rare and endangered, who could not fly very far. The golden eagles figured out quickly they could swoop down and have lunch or dinner any time they wanted.

John Fairfield didn't want to kill a golden eagle. After all, they were here first, and endangered, too. So he and his staff began to take meat up to the hill overlooking the area, place it on the rocks and let the sun cook it, and when the eagles got ready, they'd come down and eat. It worked. The golden eagles dined at leisure and left the Park birds alone. "That's how Eagle Peak got its name," he explained. "We sometimes had three or four wild eagles at a time sitting up there eating."

Such situations happened every day. "We could never count on anything happening twice," John said, then added with a laugh: "Nor did it ever enter Dr. Schroeder's mind that he was ever giving you something you couldn't do."

Life and Limb

Make no mistake, these keepers were risking a few limbs for all this new information about animal behavior, and most of them have "war stories." Randy Rieches will never forget one such moment.

"We successfully had placed black rhinos and Southern white rhinos in the same area. Then suddenly a black rhino baby was born, and it was like a light switch went on. There was a total change in dominance," explained Rieches. "The black rhino became extremely aggressive—black rhinos are 2,000 pounds to the Southern white's 4,000. The aggressive behavior got to be a daily situation while we waited to see if it would work itself out.

"One day I was feeding in the area by myself. Suddenly a very, very aggressive altercation between a white and a black took place in front of me. The white finally flipped the black rhino into the feeder and was trying to gut him—the black's whole abdomen was exposed. So I rushed the truck at them, hit the Southern white rhino head-on and pushed him backwards. He backed up and slammed into the truck three or four times, pushing me and the truck backwards, backing up and slamming, then backing up and slamming again. Finally the truck died and while I tried to start it, the Southern white rhino came around to my side, slammed the truck again and tipped the truck over.

"Funny thing was we were doing a lot of filming back then, so we could learn from it. I had the camera going, but when the rhino tipped the truck over, I dropped the camera. Let's just say the camera was still picking up some things I was saying that I didn't want taped for posterity," Randy added with a laugh. Finally, he was able to get to the radio, call for help from the other keepers, and somehow escape injury.

Rich Messina, the Park's current animal care manager, wasn't as lucky. One morning he drove out to check on the herd of Przewalksi's horses. He found a stallion called Basil attacking a newborn colt. Rich ran to the colt to get him out of harm's way when Basil bit his left forearm, crushing it. He regained the strength in this arm eventually, but the colt died.

Why would Basil kill a newborn? They discovered that the leader of a band of mares, if given the chance, will kill all colts that aren't his. The colt Basil killed had been sired by a stallion from Russia. That information is now thought to be invaluable since these horses no longer exist in the wild.

Fund-Raising, Cowboy-Style

At least, while they might be risking life and limb, those early keepers missed being drawn into fund-raising with Charlie Schroeder as John Fairfield did. Fairfield remembers well how Charlie introduced him to fund-raising, cowboy-style.

"Dr. Schroeder would call at 10 a.m. and say: 'I'm going to be out there this afternoon, and I've got some people with me.' That meant he was bringing visitors to the Park and that he wanted me to take them on a guided tour. When they arrived, he'd introduce me and jump in the back of the van, saying, 'John, put on the $5 tape.' Or it would be the $10 tape. Or the $15 tape. That was code, of course. If he said $5, that meant, 'I've got the money in my pocket, so all you need to do is just a quick run-through. If he said $10, that meant 'I think I've got it. All we need is a little more help from you.' And if he said $15, that meant 'I have drawn a blank. We are batting zero—as long as it takes, do it.' By the end of each tour, we'd be sitting up on that hill eating fried chicken, talking to these people who could buy and sell the whole place."

One time Charlie arrived with Arthur Godfrey and Mayor Pete Wilson in a zebra-striped suburban van. "I hopped in to start the tour and drove by one of the keepers who was scooping rhino dung with a skip loader," said John. "Rhinos really put it out, too. They can put 200 pounds in and get 250 out. Arthur Godfrey leaned out the window and said, 'Hey, can I do that?'

"'Sure,' I told him. I stopped the van and said to the dung-scooping keeper, 'This is Arthur Godfrey and he wants to scoop a little rhino dung.'

"Meanwhile, Pete Wilson was having a cow," said Fairfield. "'Arthur, we've got to go!' the Mayor was yelling from the car.

"Arthur turned to Wilson and said, 'Peter, why don't you just go ahead and give that speech for me. I'm having too much fun.'"

Arthur loved the Zoo and the Park, and he became one of the Zoological Society's strongest supporters. That day, John Fairfield found out why Godfrey was such a early conservationist. As he was scooping rhino dung, he saw a sable antelope. "See that?" Godfrey said to Fairfield. "That was the last thing I killed in Africa. That sable antelope is the most beautiful animal, and I shot it. I felt so bad that I bought 600 acres in Virginia for a reserve and started working with wildlife."

Wildebeest and Hospitals

John Fairfield and his growing number of keepers needed a hospital and everyone knew it, including Charlie. They were caring for the animals in a makeshift clinic. One day Dr. Schroeder called John and told him he was to give a tour to an editor and his wife from the *Escondido Times-Advocate* that afternoon—by himself. When they arrived, he was right in the middle of wildebeest calving. Dr. Les Nelson, the regular vet, was away, so John had his hands full.

John Fairfield didn't even have the zebra-striped Suburban to use. All he had was an old pickup truck with rhino dents. He put the couple into the front bench seat and asked them to bear with him. "I apologize, but we've got wildebeest cows having babies, and I gotta be there," he explained. He then left the couple alone for more than an hour while he pulled two wildebeest calves from their mothers.

"It was almost dark by the time I got back to the two in the truck. They had sat there the whole time, with no water, no restroom stops, no nothing, just watching us." He apologized to the couple, promising a better tour if they'd come out again. Their response? "It was the most wonderful day we ever had!" they both exclaimed.

On Monday, Dr. Schroeder called John, all excited. "The couple you gave the tour to—that was the son and daughter-in-law of Jerene Appleby Harnish, who owns the Escondido newspaper!" he said. "Now Mrs. Harnish is coming right down for her own tour."

Within minutes, she drove up with her chauffeur. John and Charlie escorted her to the zebra-striped Suburban and Charlie said: "John, put on the $15 tape."

During the grand tour, John explained at length about the Przewalski's horses, which Mrs. Harnish found fascinating. "We toured the whole Park," said Fairfield, "then stopped by her limousine where she pulled out a ledger, flopped it on the hood, and wrote out a check, which she handed to Charlie. 'Charlie,' she said, 'you get this to the bank before my broker finds out because he is going to pitch a fit.' Then she wrote another and said, 'This one is for John and those pretty little horses he likes.' The first check was for $100,000; the second was for $25,000."

At the dedication of the hospital named after her, Jerene Appleby Harnish explained her enthusiasm with these words: "Ecology in the truest sense of the word means the science of the relationships between organisms and their environments—a perfect definition of the San Pasqual unit of the San Diego Zoo. When the opportunity to become a part of this famous venture presented itself, I was overjoyed."

CHAPTER 23

Rhinos on a Handshake

During those pre-opening years, one of the best stories is the saga of the twenty Southern white rhinos brought all the way from South Africa by slow boat to be the first herd of the Park—a megaton experiment about the impact of space on the reproduction of this species in captivity. Their success was so dramatic that to this day the image of the Southern white rhino is our Park symbol.

The story of how they got here is fascinating.

The story begins with Ian Player, the Republic of South Africa's chief conservator. The rhinos' trek to San Pasqual Valley began that day in 1964 when Charlie took Ian up to his favorite hill where they exchanged ideas that ultimately became integrated into the Park's concept. Ian Player saw a future place for his rhinos if the Park was ever built.

"In 1970, my department in South Africa wanted to start allowing the shooting of white rhinos again in Zululand Park," Ian began. "We had been too successful in our breeding. By then, we had put in a very satisfactory gene pool of white rhino, restocked most of South Africa's parks again, as well as placed fifty around the world. They said nobody else wanted to buy them so I asked for permission to go see if I could sell them myself.

"In essence, my strategy was quite simple," Ian went on. "I would first persuade the Whipsnade Zoo in London to buy twenty rhinos, which would help me sell more elsewhere. I camped on the doorstep of the director of the London Zoological Society, who kept saying it wasn't his decision and kept putting me off." His department had given him a month. With only three days left, in a separate saga of intrigue, he finally was able to sell twenty rhinos to Whipsnade.

"I knew exactly what to do then," Player explained with a gleam in his eye. "I was on a plane the next day, straight to San Diego. I knew there would be the right kind of thinking there, and I was friendly with Charlie. I had a very clear goal. I wanted the rhinos in a place where they could breed in sufficient numbers that even if South Africa were to blow up, it would be possible to restock the country, and I wasn't too sanguine about them breeding in the cold climate of Whipsnade. San Diego, though, was the perfect place climatically."

Ian landed in San Diego, called Charlie and soon found himself eating dinner on San Diego Harbor with Charlie and Andy Borthwick. "I can still see it very clear in my mind's eye, the airplanes coming in, the harbor, Charlie and Andy and me sitting and talking," Ian remembered.

Finally Charlie turned to Ian and said, "What's your proposition?"

"Charlie always had a very funny saying: 'You've got to have a buck to run a zoo,'" recalled Player. "So I said, 'Charlie, not only do you need a buck to run a zoo, but you got to have a buck to keep the rhino alive.' I told him about the situation in South Africa, that they were wanting to begin shooting the rhino again and how I was convinced that the rhino was still not safe. Then I told him what I had sold to London. He perked up, asking how many they took. 'Twenty,' I informed him.

"'Twenty!' he exclaimed. He rapped the table with a knuckle and said, 'Then we'll take twenty.'

"Just like that," Ian said, still impressed. "No three-week wait, not a whole lot of talking. He just looked at Andy and said, 'Don't you think so, Andy?' And Andy agreed."

Charlie told Ian to send them the paperwork.

"Charlie," I said, "that's the worst thing for my situation."

Charlie thought a moment then said, "All right, we'll shake hands."

And they did.

When Ian Player flew back home, the Zululand accountants asked for the agreement. "There is no agreement," Player remembered with a wry smile. "They paled, almost became hysteric. 'Well, how have you done it?' they demanded.

"'We did it on a handshake,' I told them. This is Charlie Schroeder of the San Diego Zoo we're dealing with; a handshake is more than good enough for that man."

On January 1971, more than a year before the Park opened, Charlie Schroeder received a telegram from Player. The cable read:

> 22 WHITE RHINOS DEPARTED 0600 HOURS 20 JANUARY. LARGEST CONSIGNMENT OF RHINO EVER TO LEAVE SHORES OF SOUTH AFRICA. WE ARE HAPPY THEY ARE GOING TO YOUR RENOWNED ZOO, AND WE WILLINGLY ENTRUST THEM TO YOUR CARE.

Bringing the Rhinos Home

John Fairfield was sent over to South Africa to help capture and bring home the rhinos. His tale alone would fill a book, riding with his dart gun in a jeep while Zulu horseman wearing canvas shrouds and American football helmets—obtained from who knows where—rode ahead to track the rhinos.

But the most famous part of his rhino saga was the twenty-three days he spent on board a tramp steamer bringing them back to San Diego. After training each of the twenty-two rhinos to stay in their side-by-side crates on the boat, set strategically "so that they could see each other or else they would have gone crazy," as Fairfield explained, he also had to train them to eat and drink from buckets for the long haul.

Two of the twenty-two rhinos were tagged to go to the San Antonio Zoo, the rest to San Diego. As they started off across the Atlantic Ocean, he and Ivan Steytler, Natal Park Board senior park ranger who came along, had several tasks: feeding, watering, and doctoring the rhinos around the clock. John Fairfield will never forget that January 1971: "I left Durban, South Africa, on the same day a rocket was fired from Cape Canaveral carrying Apollo XIV. Astronauts went up, left their space ship, walked on the moon, shot a round of golf, got back in the space ship, and landed in a whole other ocean, all while I'm still trying to cross the same 'pond' back to the U.S."

He also will never forget the International Yacht Race between Cape Town and Rio de Janeiro, which was also happening at the same time. He could see the sailing yachts with his binoculars. "We were going so slow that the smoke was going faster than we were," he said. Four days at sea, they lost one of the San Antonio-bound rhinos from an infection, and had to bury it at sea, The next day, the radioman informed them that he just received an SOS from one of the racing yachts, which was just about where they buried the rhino. The SOS reported that their yacht was rammed by "a whale." The yacht sank, and the people spent nineteen hours in the water before being rescued.

John dreamed about those people for years. "Can you imagine owning a big fancy yacht that was probably hit by a dead rhino out at sea? Of course, they'd say they were hit by a whale!"

Finally, the steamer landed at Galveston Bay, where the rhinos were inspected and quarantined by U.S. custom officials. The rhinos were boarded onto Union Pacific Railway cars to San Diego. Dr. Les Nelson and two more keepers met Fairfield and Steytler in Galveston to ride in the caboose behind this trainload of rhinos. "We'd round a corner and I could look out the window and see all the rhinos," said Fairfield. At every stop, he was taken by car to the front of the mile-long freight train to feed, water and medicate the rhinos. After a detour into Mexico for safe tracks to handle the load, they finally made it into San Diego, where the rhinos were off-loaded onto flat-bed trailer trucks and hauled to the Park.

One of the most riveting early photos of the Park is of John Fairfield standing atop one of the crates as he worked to pull up its door and let the first rhino run loose after its long ordeal. In the words of Charlie Schroeder, the rhinos would have space to "run and play and do what comes naturally." Since 1971, their story has become one of the great captive breeding success stories of the modern zoo world. As of this writing, eighty-six Southern white rhinos have been born.

New Yorker writer and anthropologist Emily Hahn wrote these words after seeing the Park: "Nobody has ever before tried to create this kind of park, much less succeeded in doing so." Charlie's plan was already unfolding in surprising ways. Now it was time to open the Wild Animal Park to the public, so Charlie's grand vision could finally unfold in all its glory.

CHAPTER 24

The Zoo of the Future Opens

Finally, the great day arrived—May 10, 1972. In as natural an environment as a captive world could be, Park visitors would come see the creatures we share earth with, as Charlie believed it was important to do.

Lena, an elephant, charged through a paper bull's-eye at the visitor's gate at 11 a.m. to begin the dedication ceremonies. The first man to walk through the doors was a seventy-year-old grandfather. Waiting for him and hundreds of others were the 1,000 animals, birds, and reptiles in the Park's Nairobi Village and along the monorail ride. Carol, the Zoo's famous painting elephant, took a bath in the Park's watering hole with the aid of Joan Embery, Miss Zoofari.

Members of the Board of Trustees, government officials and dedication speakers sat under white tents on a Faust-inspired wooden bridge overlooking an African-like watering hole. Beyond the tents were 600 acres where rhinos romped, giraffes lumbered, and gnus cavorted, all in eye-shot of the people sitting under the dedication tent.

The clock in the village tower beyond the Congo River fishing hut struck noon with twelve loud bongs as Andy Borthwick was speaking—and then a sprinkler head unexpectedly showered the guests until the maintenance crew hurriedly shut it off.

"I hope you appreciate the extra shower," Andy quipped. "That water costs about twenty cents a quart to come all the way from Lake Wohlford."

Ian Player was invited to speak at the opening ceremonies. He was deeply moved when he saw his rhinos for the first time on that big day. "The rhinos are marvelous," he said. "They are doing exactly as they do in the wild. They have established small groups and have staked out their own territory." The San Pasqual's rhino herd was close to the number left in existence when Player began his campaign to save the gentle giants. "In a very real way," he remembered, still full of emotion decades later, "it was a culmination of something I'd started in 1952."

His opening remarks at the Park's ceremony said it all, though, for most of us who love the Park.

"I am a religious man," Ian said, "and my religion is of the wilderness. This Park is brilliant in its conception and execution. No one can sit on this wilderness threshold without experiencing a spiritual uplift."

Then everybody piled on board the monorail.

Betty Peach, Zoo writer for the *San Diego Evening Tribune*, described the day this way:

> The first car of the electric train, filled with the honor guests, broke through ribbons and clouds of balloons to take visitors on the five-mile continuous view of animals from various parts of the world. There were exotic, rare and endangered species—addax, eland, gnu, gazelle, giraffe, impala, kudu, lion , nilgai, rhino, springbok and zebra—some in pairs, some in families, some in herds.
>
> And mixed in with them are native chipmunks, having a feast at the daily fresh feed, bobcats, mule deer, an occasional skunk and, according to the train's commentator, a mountain lion that comes through the more remote

> regions of the Park about once a week. The Park will open every morning at nine.

And so began the San Diego Zoological Society's revolutionary venture called the San Diego Wild Animal Park.

A local newspaper editorial captured the moment in the historical context of San Diego's long, special relationship with its Zoo:

> WILD ANIMAL PARK VERY SPECIAL
>
> Shortly after the turn of century, a group of far-sighted San Diego citizens persuaded the city to issue bonds in the amount of one million and obtained private grants to sponsor the 1915-16 Panama Pacific Exposition, which was largely responsible for converting some 1,400 acres of chaparral wasteland into Balboa Park.
>
> At the turn of the current decade, the citizens of San Diego under the leadership of the Zoological Society approved a $6 million bond issue to supplement thousands of dollars of private donations to inaugurate the San Diego Wild Animal Park. Another municipal jewel has been cut and formed.
>
> Perhaps nothing will ever supersede Balboa Park as a cultural, zoological, botanical and social centerpiece of our city. Nevertheless, the Wild Animal Park certainly will challenge it in the near future.
>
> The educational and scientific aspect of the park can hardly be overstated. It will enhance ecology in the most literal definition of the word, adding to our scientific knowledge of animals that are able to roam in relative freedom. It will add to our knowledge of the relationships between the animals and human kingdoms as well as to the greater understanding of animal and plant relationships. Not the least of San Pasqual's values will be its role in preserving vanishing species.
>
> San Diegans who visit their Wild Animal Park in a sense will be pioneers. Large as the park is, they will still see only the beginning of what San Pasqual will eventually be. That will take ten to fifteen years, millions of

additional dollars and great assistance from Mother Nature.

What they will see, however is something that already is really very special—an exciting zoological tableau that already is without counterpart in the world and an undertaking of the breadth and scope that is worthy of a city in motion.

—*San Diego Union,* May 7, 1972

CHAPTER 25

The Park That Charlie Built

Picture yourself in the back of a banged-up truck riding along a rutted road straight through a crash of rhinos. *They don't call it a crash for nothing,* you think, with a glance down at the dent in your side of the truck. Here in East Africa, giraffes lope on your left, antelopes skitter off to your right. And there is North Africa across the way, and up on the hill is Asia.

Where are you?

In the middle of the San Diego Wild Animal Park.

What your eyes see is so unusual that the best we have ever come up with to describe the experience is a few words of awe: *No Place Like It on Earth.*

A writer for the *Christian Science Monitor* described her own experience eloquently:

> I never thought to see myself mixed with a herd of white rhinos, but there we all were. The only trouble with

> a rhino head is that it won't fit into a Jeep window, but it tried. Big, beautiful, nearsighted eyes were halfway down the face, far below two uptilted ears. Above the square snout was a curved horn. After peering at us for awhile, these animals decided we weren't remarkable. We watched them go play in their pond and cavort ponderously. After emerging, each felt so good that he had to cavort some more, and push his friends around.

The Spacious Ark

Over the next decade, reproduction would literally, happily explode over the open space of Charlie's Park. By 1985, the *Los Angeles Times* expressed the wonder of it all in an article about Charlie and the Park: "The roster of births include forty-eight African cheetahs, forty Przewalski's wild horses, 143 Arabian oryx, eighty Grevy's zebra, six Lowland gorillas, six Indian rhinos, more gazelles than you can shake a stick at, and fifty-nine Southern white rhinos—perhaps the Park's proudest achievement because the animal is no longer on the endangered list."

The latest figures, at the time of this printing, show that the cradle roll continues rolling on: 119 rhinos have been born at the Park—eighty-six Southern whites, twenty-eight Indians and nine East African blacks. Add 290 Arabian oryx, 112 Przewalski's horses, 117 white-tailed gnu, eighty-nine African cheetahs and sixty-six California condors, to name the some of the most endangered successes.

Dozens and dozens of other rare and endangered species round out the roll call: fourteen Lowland gorillas, 330 Abyssinian ground hornbills, 117 Grevy's zebras, 466 Scimitar-horned oryx, 323 Slender-horned gazelles, seventy-two East African bongos, and fifteen species of storks that collectively have produced 143 chicks, including the first African open-bill stork in captivity.

Today, the Park is a sanctuary for fifty-one endangered species, forty-one of which have successfully reproduced there. Home to 3,200 rare and endangered animals. Another 500 to 600 animals arrive by birth or hatching each year.

Looking out over the Park one day in the mid-eighties, Charlie remembered a day in Rome. "I was there for the International

Union of Zoo Directors. I remember walking through the zoo and seeing a white-tailed gnu and thinking, *Oh, boy, I would give anything for a white-tailed gnu.* And what do we have now? A herd of fifty? If anybody'd told me then I'd have fifty, I wouldn't have believed them."

In fulfilling Charlie's first mission for the Park, today the Park ships cranes to Hong Kong, cheetahs to Oregon, okapi to Amsterdam, Arabian oryx to Oman, and Andean condors to Columbia. Charlie's Park is accomplishing its original goal.

"He's populated the world's zoo with exotic animals," said Jim Dolan. "There is no need to go to the wild anymore. We are a remarkable institution. We don't do much selling. Instead, we have on loan to other institutions an enormous number of animals out of the Zoo and the Park. In terms of the conservation ethic of this institution, that is an important issue—that the collection does not represent financial gain but instead a philosophical gain. We work with other zoos, sometimes housing their endangered species group, in order to refresh bloodlines."

The first thirty years at the Park have seen many firsts and many surprises—the prodigious Southern white rhino reproduction, the biggest gorilla exhibit in the zoological world, the first Abyssinian ground hornbill chick hatched in the U.S. and the second in captivity anywhere, and the first African cheetah birth to survive in captivity since the fifties.

But as Charlie would say, we didn't do anything magical. It was just lots of space and warm, natural environment. Perhaps that's what the real magic was from Charlie's view.

Of course, even Charlie could not see some of the remarkable ramifications of his loyalty to his dream, including the explosion of knowledge the Park would offer research—his first love. "A grand vision brings all these things with it," said Andy Phillips, CRES deputy director. "The Society's research department couldn't have quadrupled without the Wild Animal Park, for instance. Regardless of whether Charlie foresaw the potential for this kind of growth, the jump in knowledge and the increased staff are there because of him."

As mentioned earlier, the Southern white rhino saga is the Park's quintessential success story. Three other breeding efforts are standouts—the Californian condor, the African cheetah, and the Arabian oryx. These successes are the stuff that made Charlie grin that smile of his and say, "Everything is just unbelievable, things I never dreamed of, never thought we could do."

California Condor

Thirty years ago, the California condor was going the way of the dodo, and everyone agreed something should be done about it. But the groups most concerned could not agree what that something should be. The debate lasted several decades. Back as far as 1953, the controversy was making news.

Most everyone believed that the major reason the giant, regal vulture was dying off was because of the loss of its habitat to California's incredible human population growth. The suggestion to bring the remaining birds into a safe place to breed met with well-meaning resistance by groups such as the National Audubon Society, which feared if we took the birds out of the remaining habitat, then the habitat would be lost forever to development.

As the years passed, the debate continued to rage and the condor numbers continued to dwindle.

By the early eighties, there were only twenty or so condors left in the wild. "They were dying off at an incredible rate for a variety of reasons," explained Mike Mace, the Park's curator of birds. "Keep in mind that this is a species that occurred in the Pleistocene Era. Lewis and Clark saw these birds. Their historic range is the entire West Coast, all the way through the southern portion of the country, and up into New York. The decision was made—a very difficult decision in those days—to bring all the birds in for protection and propagation, and we were chosen because we'd had such good luck breeding the Andean condors at the Zoo." But that didn't stop the lawsuits filed by those who worried about the loss of the bird's habitat.

The good news is that bringing in the remaining birds was the right decision, and the California Condor Recovery Project has become one of the centerpieces for recovery programs of endangered species. Nearly half the California condors alive today were hatched at the San Diego Wild Animal Park. These successes have allowed scientists to begin cautionary release of captive-produced birds into their native habitat in 1991. The Park continues its efforts to re-establish North America's largest flying bird in the wild in partnership with the Los Angeles Zoo and the Peregrine Fund in Boise, Idaho. In 1997, condors were released in Arizona in the Vermillion Cliffs area north of the Grand Canyon and in the Big Sur areas of California. The talk is of taking them to other parts of the country that have fossilized record of the condors.

Although everyone knows the condors are residents of the Park, in their fifteen years of residence on the hill in their "condorminiums," they have never been exhibited. Only now are plans being suggested to incorporate a limited exposure of the condors to Park visitors. Their seclusion reflects the original intent of the Park—to establish a breeding facility best suited to reintroduction. The species needs a healthy fear of humans, which is why when condor chicks are hatched at the Park, they literally never see a human being.

"We take great pains when raising California condors to 'puppet-raise' them for the purpose of not letting them imprint on human beings, to keep them from seeing humans as the literal hand that feeds them," explained Mace. Condor chicks are fed with condor mother puppets, after which they are grabbed in the dark and whisked away in shrouded kennel cages to be taken to the Los Angeles Zoo for their socialization, the first step to reintroduction. Today, the cooperative program thrives and condor breeding at the Park continues to be one of our proudest accomplishments.

Cheetah

Five pair of African cheetahs were brought to the park prior to its opening through a research program with the Donner Foundation.

John Fairfield took an instant liking to them, which turned out to be not only fortunate for the cheetahs, but also for the research. Fairfield could look at an animal and see subtle changes. One day, one of the female cheetahs looked different to him; he could see its nipples showing. For several days, he watched the female. Finally he called the designated Park veterinarian Les Nelson, still headquartered at the Zoo and said, "Doc, I got a cheetah here who's just told me that she's gonna have some kittens. I can't tell you when, but I feel pretty certain."

What makes this prediction even more interesting is that he was very aware that only one cheetah had been born in captivity in the world since the fifties. He also had a hunch he didn't much like. He suggested that Nelson allow them to move the males out of the cage because he suddenly feared, like a tomcat known to kill baby kittens, that the males might do the same. As he put it, "I'd sure hate to have kittens born and not be able to raise them."

Before they could decide, however, John was in for a surprise. Around noon, something told him to go up to the cheetah cages. What he saw just inside the six-foot double fence took his breath away. There sat the female cheetah with three little bouncing baby cheetahs, still wet . . . and sneaking up the hill toward her was a big male. Right in front of his eyes, the male cat grabbed one of the babies and killed it. Before John could even think, the male cheetah had grabbed up the second one and killed it, too.

There was only one cub left, and the mother cheetah was trying to fight off the male. John instinctively scrambled over the double fence, grabbed a sumac brush limb, shoved it in the mother's face, scooped up the kitten, shoved it inside his shirt, and sprinted for the fence, adrenaline pumping. He rushed down to the security officer's desk in the little three-walled hut and said, "Call the San Diego Zoo. Tell them I have an important package, and I would like for them to meet me at the back gate." The vets came, their ambulance siren wailing.

"What's the big deal?" one asked.

Fairfield pulled the cheetah cub from his shirt and handed it to him.

They named her Juba.

At that time, December 1970, Les Nelson, the Park's veterinarian, called Juba, "the most important birth in the history of the Zoo."

One of the wonders for early Park keepers and their species reproduction goal was the impact of pure observation, much like Fairfield's intuitive rescue that saved Juba's life. It was certainly a dynamic that Charlie Schroeder must have found fascinating, as did others in the zoological world.

"Male and female cheetahs, we realized, cannot be together constantly and reproduce," Park pathologist Dr. Lynn Griner explained two years later when the second litter from Juba's mother was born with the help of John Fairfield's experience. The keepers began isolating the males and females from each other in October 1971. The question became, *When do we put them back together?* The keepers reintroduced them in January 1972 when they noticed that males were beginning to build mounds of dirt.

"We tried for months to determine when the females were in season without success. The mound-building by the males has been our single-most important clue as to when the introduction of males and females should take place. Human observation couldn't detect the female's coming into season, but the males could tell by scent," Dr. Griner said.

That was the way—and is still the way—that the miracles at the Park happen—through what observation teaches us.

While the off-sight research and breeding spaces continue, a few cheetahs are now in their new home on a green, moated hill overlooking the entire African and Asian wide-open exhibits, visible, now and then, by fortunate visitors on the Heart of Africa trail, the latest addition to the Park. "Any closer, you'd be lunch," states a billboard advertisement announcing the new "up close and personal" experience.

Arabian Oryx

Think of the unicorn. Some believe that the Arabian oryx inspired the myth. The oryx is a beautiful desert creature with horns that look so perfectly identical from the side that it's easy to see how they might have been the inspiration for such the wonderful imaginary creature. The Arabian oryx once ranged in herds over the entire Arabian peninsula and into Jordan, Syria, and Iraq. The herds were decimated by hunters, often guests of the sheiks who chased them in Jeeps with high-powered rifles for sport. Add all the desert tribesmen who hunted them, believing that the man able to bring down the majestic oryx would rightfully inherit its courage and strength. When they began to use the less-than-courageous help of new high-powered weapons, the oryx had little chance.

Fewer than 200 Arabian oryx were believed to exist in the world when the oryx came to the Wild Animal Park from the thirty "world pool" placed in the Phoenix Zoo. They produced so prodigiously at the Park that the Arabian oryx became our first attempt to reintroduce a species back into it original habitat.

"The first oryx to go back into the wild is probably the fondest memory I have," stated Jim Dolan. "That's what we look forward to. We can only hope there will come a day when we can send any endangered species we're breeding back to where they originated. Along with other institutions, we were able to put this reintroduction program together. Fortunately the Sultan of Oman was very interested in having them back, and put lot of his own money into the project."

More than a hundred were put back into the wild in Oman, but over forty were soon poached out. That didn't stunt Jim Dolan's enthusiasm for the effort. "There's never a guarantee," he said. "The Sultan of Oman may request more animals for us, so thank God we have this captive population that continues to grow."

To talk about the Park as nurturing conservation projects, this has to be "the star of the crown," as Jim Dolan has put it. "The Arabian oryx was an animal totally exterminated in the wild, which has now been put back in the wild with offspring of captive-bred animals. "Whether this is possible with a large number of species remains to be seen," Jim admits. "This is the quandary the world faces."

Beyond the Ark—Conservation and Habitat

What most of us find fascinating is that only now are the world's zoos embracing their vital roles in conservation, understanding the nuances and needs, and how we as institutions can meet them. Yet because of Dr. Schroeder's uncanny far-sighted vision, we already have what most of us would only now be dreaming of.

"I think when Charlie created the Park, he smiled and said, 'There! Now what are you going to do?'" remarked Bill Toone, director of applied conservation programs and former Park curator. "He challenged us to take the next step, and we've taken up that challenge. In doing that, we're breaking a centuries-old mold of what zoos are all about. This is probably an honest tribute to Charlie's vision that thirty years after creating the Wild Animal Park, most zoos have yet to catch up. The Wild Animal Park is still, today, the zoo of the future, so that puts us in a wonderful position to plan the next one."

Because of Charlie Schroeder, we are more than a step ahead. We are represented on five continents with dozens of conservation projects, we're constantly working on new reintroduction programs of endangered animals bred at the Zoo and the Park, and we have the Applied Conservation Programs with its conservation curator and habitat analyst who works with environmental, educational and private agencies to educate the citizenry about the need for conservation of habitat, including local ecosytems.

We participate in worldwide communal projects as well, including the International Okapi Breeding Project which works to build that prized species' numbers in captivity. The curators also work closely with the International Species Inventory System and AAZPA's Species Survival Plan (SSP) established to help avoid a Noah's Ark approach to zoology, with two animals here and two animals there, reproducing haphazardly.

The Species Survival Plan (SSP) monitors and recommends specific breeding loans and gene pools. The curators use the system to learn where the best genetically suitable mates can be found for particular animals, and they ship animals to places as far away as Czechoslovakia, as in the case recently of a male Northern white rhino, to keep the breeding pools strong in the species we can save.

CRES deputy director Andy Phillips once explained vividly why such conservation is fundamental: "On a scale of 1 to 10 in keeping the world wildlife viable, with 10 being the best, we are at a 3. We can't get back to a 5, but maybe we can get back to 4. All we truly want to do is never get to a 1, where there isn't anything left in the wild anymore. There are so many species in the world dying off yearly, but there aren't enough scientists to work on even a hundredth of them. It's unfortunate that scientists have to play God, but we don't have enough time to work on everything, so we must choose the ones we are going to work on."

Trotting Around the Ark

Everything is interrelated, as Charlie would have been the first to explain. The quality of our life on earth is affected when species disappear in more ways than we can know. All we know is once a species is gone, it's gone forever. Bill Toone vividly ponders one sure way letting species die will affect us: "What happens if we allow an endangered species to go extinct? Nothing. We can continue to allow it and allow it. Our planet will remain full, but maybe the birds sharing the place with us would all be house sparrows and that would be terribly, terribly dull."

Species going extinct is a part of nature, of course. The present habitat loss and species loss is far beyond the natural rate, though, and is now, head-spinning in its acceleration. To save an endangered animal, then, is not only to bring them to the Zoo and the Park and keep them breeding, but to go out and try to keep other species from becoming endangered too, by saving their habitats.

"People tell me that animals would be better off in the wild, and I say, 'What wild?'" CRES director Werner Heuschele said so eloquently, and those words are far too prophetic for comfort. Yet in them lies our future work. Not only do we do things here, we do things in the field. Our involvement increases every year, completing the circle, one with the other. Otherwise, captivity becomes an island when there should be interplay with the captive and the wild.

Jim Dolan explained it further: "It's not just animals we are interested in now. All our work, our endangered breeding programs, our reintroduction dreams, is dependent on whether the environment is still there, so environment becomes an issue for all zoological institutions. To educate the public in understanding the need for wild places is now the major thrust of any zoological institution. We are not only saving one animal when we save the habitat. We are saving the whole 'biosystem.'" Then Dolan put the problem as personally as most of us feel it: "It troubles me that there are things I will never see because people got greedy."

Our mission now includes responsibility outside the Park, as well as the responsibility of an essential ark for a world with a shrinking wild.

"Charlie never meant for the Park to be the only wild for these creatures," Jim added. "But the spiraling loss of the world's wild habitats happened so fast, too fast. In reality, we are now more like arks than we care to be. With the loss of habitat, pandas, gnus, and far too many others are trotting around the edges of our lifeboat waiting until they have some place to go. This is why conservation and habitat efforts must now be a zoo's main concern. We have no choice. Charlie knew that."

Bill Toone describes the situation even more dramatically: "If all you're going to do is save an animal in a zoo, that's called preservation. That's taking a genetic resource, putting it in a bottle, putting a lid on it, and saving it. In a hundred years, someone can come to a zoo and see a giraffe. But conservation is making sure that some child can have the same experience I had when I went to Africa and saw a giraffe browsing. The same as going to New Guinea and see birds of paradise in the canopy, traveling to Madagascar and watching lemurs float through the trees and bark at you. The only thing that keeps the lemurs in the trees is the fact that there's a tree to jump into. The only thing that keeps giraffes on the savannah is the tree to browse in and grasses to walk through. Those are the habitats—the functioning ecosystem we now are working to save."

While we work at the captive breeding concept that Charlie started, we also work to re-establish wild places, to store endangered animals' genetic material while we work together to keep the gene pool healthy and, as Jim Dolan expresses it, "to hold our breaths to see if our efforts will win the day."

What Would Dr. Harry Think?

So what of the future?

"If you had said to Harry Wegeforth, 'What we need to do is fence in 1,800 acres in San Pasqual and turn all of these things loose,' he'd have looked at you like you were a moron," said Bill Toone. "But today, all the zoos are saying, 'How do we get 1,800 acres and turn animals loose?' The next challenge may sound just as ridiculous and to say it involves the whole world might sound insane to most people. I hope we honor Charlie's spirit by being open to what the future tells us we need to do, just as it did for Charlie Schroeder."

We're pushing on to the next generation. We believe that we can make a difference because of Charlie. It was his best lesson for us. How we do things for the future of a species or plant or animal habitat is the question. Who knows? The answers may come from surprising sources. A good example is the existence of Dr. Kurt Benirschke's Frozen Zoo and the rest of the efforts of the Center for Reproduction of Endangered Species, which are having more and more of an impact on an uncertain animal future. A quote by Librarian of Congress Daniel J. Boorstin posted above the Frozen Zoo states the unsettling but responsible situation far too well: "You must collect things for reasons you don't yet understand."

The Park, like Dr. Harry's Zoo almost a century ago, has turned into an incredible, enlightened place, a touchstone for other zoos as we all fight the encroaching problems of the endangered animal world. As for the experience it offers anyone who walks through the front gates, a visitor would have to travel 5,000 miles to see anything comparable. "The Park has drawn a lot of admiration. It's really hard to duplicate creation, but we're as close as you're going to get in an artificial environment," remarked Jim Dolan. Wherever finances are available today, most institutions would do what Charlie did here.

Charlie's Gift of Magic

Years ago, Bob Smith wrote an insightful ZOONOOZ article about the on-going love affair between the Zoo and its visitors. What has been the magic? It's more than entertainment. "Borrowing a concept from master-planner Richard Neutra," wrote Smith, "the San Diego Zoo has become a psychotrope; that is, a place that people see or hear about and then feel they want to be in and be part of."

That definition rings wonderfully true. If not, a spirit-lifting stroll into the Zoo or the Park would change your mind. While visiting San Diego, Ian Player explained the phenomenon this way: "For children confined to cities all over the world—and this is the vast majority of all children—going to the zoo is their first and often only taste of wilderness." Charlie Schroeder's mind, heart, soul, and most of the years of his adult life were given over to making the San Diego Zoo exactly that.

The San Diego Wild Animal Park, though, is that extra step, a taste of wilderness that comes as close to duplicating creation as anything in the zoo world today. That connection to creation is found daily at Charlie's Park, and it is Charlie's enduring gift to us all.

CHAPTER 26

Retired But Not Retiring

"This is still a hell of a joint, isn't it!"
—Charlie Schroeder on the Wild Animal Park
Los Angeles Times, 1985

Three months after the Park opened, Charlie Schroeder, seventy-one-years old, retired as director of the San Diego Zoo and ultimately began a new career as a popular consultant to the rest of the zoological world.

Of course, Charlie never really retired when it came to the Wild Animal Park. He was always visiting and always asking the right questions, sometimes to our chagrin. No longer was he handing out Schroeder snowflakes, but he turned his formidable charm onto the task of a gently nudging us about what his eyes still did not miss.

General curator Larry Killmar remembers visits from Charlie vividly.

"Dr. Schroeder would walk into my office and say: 'Larry, I think you ought to think about doing this or that,'" recalls Killmar, "and I'd answer with something like, 'That's a good idea, Dr. Schroeder, but we want to do it this way.'

"'You know best,' he'd answer. But what he'd really done is toss a hand grenade in my lap and by God, I better have an answer the next time he came in."

For the last years of Charlie's life, Bob McClure, current general manager of the Park, became good friends with Charlie, seeing him on his regular visits.

"At least once a week, he'd come into my office," McClure remembered. "He never asked idle questions. They'd be about future plans, how we were going to get from here to there, and how it was going to be funded. There was always the follow-up question that got to the heart of the matter. Of course, he'd also throw in at least two or three—if not ten or twelve—comments like, 'You know, you have a fence line over there that should be repaired,' or 'Your bench in the front of the restroom should be painted better because so many people sit on them.' He never missed anything. I actually started my own 'tickler file' out of self-defense," McClure said with a wry smile. "But the wonderful thing about Charlie when I knew him during his retirement years was that he was more a guide than a boss."

The Zoo's new director Chuck Bieler also would receive "Charlie" visits. "He'd say, 'Now, I want to tell you something. You don't have to agree, but you have to listen,'" remembers Chuck.

To the end, he was pushing ideas that seemed unfathomable to everyone but Dr. Schroeder—more land at the park, another monorail, and other futuristic ideas. "One of my jobs," he would say, "is to push it."

Weather Station

Even some of Charlie's ideas that we thought were wrong, we finally decided were not. For instance, he had set up a weather station at the Park to gain television exposure. Weather broadcasts on the local news would say: "And at the Wild Animal Park today, it was eighty-nine degrees."

For the longest time, Bob McClure and I tried to persuade the stations not to *always* mention the temperature, especially when it was really hot. Showing triple-digit temperatures seemed a negative for summer crowds. But finally we said, "You know what? Everybody knows it's hot. Who cares? We have our name on TV, and Dr. Schroeder was right. Let's leave it alone."

As Charlie grew older, his time at the Park became even more precious. The older employees began to make sure that the new staff knew about this frequent visitor. "Everyone at the Park knew who he was, but as new employees began to be hired, those of us who knew all he'd done for the Zoo and Park tried to explain how important the man they'd just met really was," Bob remembered. "We would take pains to explain to new staff that, if they stayed in this business, someday somebody was going to mention Dr. Schroeder in their presence, and they'd be able to say, 'Yes, I met him.'"

A Train for You, Too

Chris Di Sabato was the operations manager at the Wild Animal Park during those years, and he reminded me of his memorable first encounter with Charlie Schroeder.

One weekend night, at the end of a very long day, he was talking to Bob McClure and me when he noticed a late group coming in. Charlie had walked in with some friends, wanting to give them a tour, and I introduced Dr. Schroeder to Chris. The monorail was already closed, but since Dr. Schroeder and his guests wanted to ride, Chris had to authorize overtime for the monorail and garage crews, something he did not want to do because of the expense.

"I remember saying something about all the extra effort and cost, and your response was memorable," Chris said to me. "You said, 'When you have accomplished what Charlie has in your career, we'll take the extra effort to pull a train out for you, too.'"

Chris also remembered being befriended by Charlie after that incident. "Many times Charlie came down to the Park on Sunday afternoons to look around," he said. "We would cruise around the Park in a cart just looking and talking. He was full of questions and stories, and he would reminisce about the Society, the Zoo, and the Park. His love for the institution was apparent, as was his pride and delight in what had been built. His stories were often about the people who worked for him and what they had accomplished, though, rather than about himself. He gave me a sense of being a part of something historic and continuing. It was a privilege to know him."

Fried-Chicken Sundays

As the years passed, Charlie continued his visits once or twice a week. On Sundays, he'd come in with a bucket of fried chicken and sit down with employees, staff, keepers, and others who worked in his beloved Park. On Wednesdays, he'd bring in his mail, much as he'd done for years at the Zoo, and open it surrounded by young, new employees.

"He was such an icon at the Wild Animal Park," said longtime employee Audrey Young. "On the Wednesdays he came to the park with his mail, the privilege of being chosen to help him was great. He'd come into the employee lounge during lunch and hold court. He was so loved. We couldn't get enough of him."

Of course, Charlie was still known to size up an employee every now and then and say, "You need a haircut," but his lunches with the staff were really about his love for the place. "He'd rather pull up his chair with the gardeners and ask how the plant collection was doing than spend all his time with management inside offices," recalled Bill Toone. "Not that he was going to do anything about what he heard, but this was the place he loved, and he wanted to know how its heart was beating and how the blood was flowing. As a veterinarian, that's the way he looked at things. I think that's why he was terribly successful. He always went to the heart of what made something work."

Still Bringing People Together

I continued to be awed by his great ability to bring people together, a talent he never quit using. Several times during my years at the Park, he would pop by the office, ask if he could interrupt for just a minute and say, "There is someone in the hall I want you to meet." And in would stroll someone like Marlin Perkins of *Wild Kingdom.*

Bill Toone tells a similar story. "He'd bring someone like Gerald Durrell of the Jersey Wildlife Preservation Trust into my office, introduce us, and say, 'You know, you two ought to do something together.'"

We were all amazed by his ability, even late in his life, to make things happen. Chuck Bieler remembers that he was a hard act to follow. "When I was director, I met Gene Trepte at the Park," Chuck explained. "We wanted to build a food stand out near the hiking trail. He was chairman of that committee, so I said, 'Gene, we want to build the food stand out there by the Bird of Prey amphitheater.'"

Gene answered, "Don't give me, 'Just out there.' Let's go and look where."

So Chuck took him to the spot.

"Well, *exactly* where?" Gene asked.

"Gene," Bieler answered, "Charlie Schroeder stood on this hillside and said, 'I'm going to put East Africa there, South Africa there, the parking lot here, and the Village there, and you bought it. I want to put in a simple little food stand, and you want to know the exact damn spot."

Eugene just laughed, Chuck remembered. "He knew it was true."

Key to Immortality

Chuck remembers one more poignant impression that Charlie left with all of us. "I always thought Charlie Schroeder had the key to immortality," he said. "I felt he would go on forever because he never slowed down. He kept that same vitality, that same gleam in his eye all his days."

Charlie never quit caring, never quit challenging, never quit trying to make things better for all we were doing. It seemed his special version of the fountain of youth.

Still a Hell of a Joint

In 1985, for a *Los Angeles Times* profile of Dr. Schroeder, the reporter rode the monorail with Charlie and gave a wonderful snapshot of the man:

> The old man kept sounding off like a cynical tourist as he walked around the Wild Animal Park. "See that crack in the pavement? They oughta fix it. The food should be better, and they should serve it quicker. The line for the monorail is too long. They should run more trains to keep the wait down."
>
> Even during the ride, as the guide pointed out beautiful

> animals that are extinct in the wild, the man complained. "Look at the dead bamboo. It looks bad. They ought to rip it up. The windshield's dirty. Don't they know how easy it is to clean it in the morning? Heck, they've got the tools."
>
> But the old man knew his stuff about animals, too. When someone asked the tour guide a question she couldn't answer, she turned to him for the answer. And when the two-car train pulled into its unloading dock and the guide wished everyone a nice visit, the man turned decidedly upbeat. He applauded and complimented the guide for her work.
>
> And as he walked away, he chirped, "This is still one hell of a joint, isn't it?" Meet Dr. Schroeder. Call him Doc. Call him Charlie. Call him the man who built the San Diego Wild Animal Park. . . .

Here at the beginning of the 21st century, Charlie is still everywhere in the Park for all of us who knew him and who work with his legacy everyday.

It has always been that way and always will be. The Park is Charlie and Charlie is the Park.

At the dedication ceremony for the monorail, Anderson Borthwick's assessment of that meaningful truth is as perfect a benediction to Charlie's life, Charlie's genius, and Charlie's heart as I or anyone could write.

These were Andy's words:

"When the time comes for Charlie Schroeder to leave this earth, I'm sure his spirit will still be hovering over the San Diego Wild Animal Park as long as it exists."

Epilogue

Mister Zoo

"I am glad to see you come back to a zoo that is young for I feel you will be able to see it in full bloom before you pass out of the picture."
—Dr. Harry Wegeforth, in a letter to Dr. Charlie Schroeder upon his return as San Diego Zoo veterinarian in 1939

"It was fun." —Dr. Charles Schroeder, director emeritus, 1983

On a chilly night in December, a year before he retired, Charlie stood on a podium in the heart of the city. He had just been named Mr. San Diego of 1972, while a prized pygmy chimpanzee dressed in a red Christmas suit sat in Maxine Schroeder's lap, eating bananas and Charlie's apple pie. Charlie, true to his nature, told a story.

One of his many unforgettable moments as a zoo vet, he told the crowd, was a moment during surgery on a very big zoo patient. "I underjudged the amount of anesthetic and suddenly found myself face to face in a very small cage with a wide-awake grizzly bear. Yet that is no more frightening than being here today," he quipped.

His acceptance speech ran about two minutes and drew a standing ovation and prolonged applause—"because of the short speech," he said afterwards.

Charlie was wrong, of course. That night was just the beginning of many such moments of deserved ovation.

Charlie often dismissed discussion of his numerous awards, honors and accolades by saying, "They're really for longevity." But as *ZOONOOZ* editor Marjorie Shaw wrote, "He began reaping them as too young a man for that to be true."

On another honored night when Charlie was named Distinguished Alumni of Washington State's University's Veterinary School, Dean L. K. Bustad put it perfectly: "Charlie has been on a high lope all his life. A brief review of his career will convince you of this. It will also impress you—and probably tire you out."

The list of honors and activities and accomplishments of this man is enough to fill up several lives, but if one of those many honors has to be described as *the* moment for Dr. Schroeder, it would probably be this one:

On January 10, 1973, more than 700 people packed into San Diego's Town and Country Hotel Convention Center to honor the retirement of Charles Robbins Schroeder, D.V.M. after his nineteen years as director of the Zoo and a total of twenty-six years service to the Zoological Society.

It was one of those testimonial dinners that allows for many voices to extol the virtues and foibles of the honoree, and many voices were heard that evening.

Governor Ronald Reagan sent a commendation which read that the Zoo under Charlie's stewardship had "grown in size and stature to the greatest animal collection in the world."

State Senator Jack Schrade, representing the California Senate, called Dr. Schroeder, "a scientist, educator, leader, innovator, administrator and citizen who conveys the pride and gratitude of all California." Society's president General Victor Krulak began his remarks by saying: "The key words—creativity, dedication, and generosity—are cardinal in your makeup. . . ."

Mayor Pete Wilson presented a proclamation from the city: "On behalf of my colleagues in the city council and the people of all San Diego, we give this token of our affection and esteem because the world's greatest zoo and the world's greatest animal park happen to be in the city of San Diego."

Aardvark President Herbert Kunzel explained Charlie's fund-raising wiles: "He knows more ways of getting money out of people than you would believe. He has a hospital there where the animals get better care than you and I do."

Joining in was Leonard Goss, Cleveland Zoo director, warmly describing Charlie as a "man who makes everything live, makes everyone want to belong." He also read a warm cable from the

International Union of Zoo Directors and the American Association of Zoological Parks and Aquariums, both of which Charlie had led as president. "He is an eternal optimist," Goss added, "We can add him to the list of vanishing species."

More than a hundred letters and accolades came from all over the world to add to the night. Then former Board president Andy Borthwick used a well-known verse to express his own sentiments. "Perhaps the best way to summarize our honored guest tonight is to describe a successful man," Andy said, and quoted this anonymous old verse:

> He has achieved success
> who has lived well, laughed often and loved much
> who has gained the respect of intelligent men and the
> love of little children
> who has filled his niche and accomplished his task
> who has left the world better than he found it; whether
> by an improved poppy, a perfect poem, or a
> rescued soul
> who has never lacked appreciation of the earth's beauty
> or failed to express it
> who has always looked for the best in other and given the
> best he had
> whose life was an inspiration
> whose memory is a benediction.

Towering Giant

After that night, General Victor Krulak wrote Charlie these eloquent sentiments: "The calendar says that today you retire from what is certainly the most distinguished zoo career in modern times. The calendar is technically correct but in practical terms, I do not believe it could possibly be more wrong. Because of your skill, your knowledge and your dedication to the whole science, it will never be possible for you to retire."

His words were prophetic.

In the years to come, Charlie would become "Mister Zoo" to all who knew him or knew of him. As Los Angeles Zoo director Dr. Warren Thomas was quoted as saying in the *Los Angeles Times,* "If ever there were a towering giant in the zoo world, it is Charlie

Schroeder." Between the banquet night in 1973 and the day he died almost two decades later, Charlie would prove just that by jetting around the globe dispensing his expertise and spreading his creative zoo genius to all his zoo friends and colleagues.

His wife, Maxine, can be thanked for inspiring him to begin dispensing his "retirement wisdom." One day, very shortly after his retirement, they were both in the kitchen. Still obviously in his longtime memo mode, Charlie looked around and said, "Maxine, there's dust on the top of the refrigerator."

"Charles," said the five-foot, two-inch Maxine, "I'd have to get up on a stool to see it. I think it's time you began to look into doing some consulting and sitting on boards or something."

That was good advice for Charlie and a boon for the rest of us. In the years ahead, in between visits to the Park and raising avocados, grapefruits, and oranges on his two-acre hilltop lot, he would fly around the world as a zoo consultant to countries such as Iran, Singapore and Israel. He would serve actively on many boards and organizations. He would also act as interim director at several zoos in the U.S., including Los Angeles, Riverbanks in South Carolina, and Audubon Zoo in New Orleans.

I first met Charlie in 1974. Charlie headed up a committee to work with the federal government's forthcoming Endangered Species Act. I had just started my career in Los Angeles and was asked if I'd like to go to a meeting with the political action committee of the AAZPA. I'll never forget walking into a board room at Sea World for the meeting. Who was sitting at the head of the table but Dr. Charles Schroeder, a man I had heard was a living legend. It wasn't even his board room, but he was without doubt in charge.

Charlie also continued to work for the city he loved. Bob Smith remembers asking Charlie to join him on trips across the country promoting the city for the San Diego Convention and Visitor Bureau. "Charlie was so fun," remembers Bob. "He was the star in every television, radio and newspaper interview we did. The media loved him."

Charlie also remained his charming scientist self, too, as witnessed by another anecdote from Smith: "One day we were in Calgary with our four children in a hotel coffee shop, and in walks Charlie, who was there visiting the Calgary Zoo. He walks over to our table, looks at our younger daughter and says, 'I knew you when you were an embryo.'"

Ron Foreman, director of New Orleans' Audubon Zoo, remembered how Charlie's talent for asking the right questions and charming the right people was a godsend.

"We brought in Charlie to be the guiding force to change the zoo from being run publicly to being a private corporation—public support and private money. I was so impressed with how he was always motivational and serious in a fun, nonthreatening way. He was a tremendous talent. He helped us turn an animal ghetto to an urban Eden," Foreman said.

Everybody seemed to know Charlie during his retirement years. "One time, I was flying somewhere when I learned the person nearest me was an architect designing a zoo," recalled Dean L. K. Bustad. "He told me the person who was giving him the best advice was Charlie Schroeder. That didn't surprise me. Many of us reached the same conclusion—that Charles R. Schroeder had more knowledge about zoos than anyone else. Wherever I go—Africa, Europe, Australia—I meet somebody who knew Charlie."

There are so many such stories, too many to tell them all. Two moments in 1986, though, were special expressions of Charlie's life and legacy:

On May 7, 1986, Dr. Charlie Schroeder was given national recognition when he was honored in a series of CBS-TV minute-long profiles called *American Portrait* that highlighted outstanding contributions to American life.

Then on October 30, 1986, a long-awaited gorilla baby was born at the Wild Animal Park. We named it Schroeder.

His Memory a Benediction

Charlie died in 1990 at the age of eighty-nine.

Up until two months before his death, he was still visiting the Wild Animal Park every week.

"They might take Charlie out of the Zoo, but they could never take the Zoo out of Charlie," a zoo colleague said at the time of Charlie's retirement. Death may have taken Charlie out of this life, but Charlie's legacy will never be taken out of the San Diego Zoo and Wild Animal Park. We continue his work, and he is with us in spirit as we do.

On his death, the *San Diego Union* wrote:

> Charles Robbins Schroeder was the man most responsible for transforming the Zoo from a charming regional animal exhibit into a world-renowned attraction. After replacing ugly barred cages with natural-looking animal enclosures, he installed the Children's Zoo and the Skyfari aerial tram—novelties that were strongly opposed at the time by some board members. Observing the successful operation elsewhere of animal refuges open to visitors, Schroeder parlayed the Zoo's need for a reserve into the San Diego Wild Animal Park at San Pasqual. It and the Zoo are models that other zoological organizations seek to duplicate. That and the esteem accorded the Zoo are tribute to the enduring work of Dr. Schroeder, who died this week at age eighty-nine.

It Was Fun

"He gets a kick from all of life," an interviewer once wrote about him, "and when he talks about the nineteen years he spent running the Zoo and his battle to build the Wild Animal Park, he sums it all up in three words: 'It was fun.'" Charlie Schroeder knew how to have fun, and he knew how to make important things like conservation and education enjoyable pursuits as well. No one will ever know how many people have been affected, how many opinions influenced, how many lives changed, and how many animals saved because he was able to make his vision everybody's vision.

How did he do it? Because he made it sound worthy and because he made it sound fun. Charlie left us all with a belief that we can make a difference. As curator Carmi Penny expressed it, "Nothing is impossible—that was the single greatest thing he conveyed to us all."

Dr. Charles Robbins Schroeder was an amazing man, the kind who comes along once in a generation, the kind who touches millions in his lifetime. The zoological world is the better because he lived. But I dare say, considering the on-going work made possible for the creatures of our earth by his hands-on dreaming, the wide world itself is also a better place for his having come our way.

Index